POPULAT  OL

POPULATION CONTROL

How Corporate Owners
Are Killing Us

JIM MARRS

wm
WILLIAM MORROW
An Imprint of HarperCollins *Publishers*

HarperCollins books may be purchased for educational, business, or sales
promotional use. For information please e-mail the Special Markets Department at SPsales@harpercollins.com.

A hardcover edition of this book was published in 2015 by William Morrow,
an imprint of HarperCollins Publishers.

FIRST WILLIAM MORROW PAPERBACK EDITION PUBLISHED 2016.

Designed by Jamie Lynn Kerner

Library of Congress Cataloging-in-Publication Data has been applied for.

ISBN 978-0-06-235990-2

24 25 26 27 28 LBC 13 12 11 10 9

CONTENTS

INTRODUCTION

ONCE UPON A TIME A GREAT NATION BECAME A SELF-DESTRUCTIVE society. Most people throughout history have been happy just to be alive. They have sought, in the words of the Declaration of Independence, "life, liberty and the pursuit of happiness."

Many Americans now recognize how far we have strayed from this ideal.

America is experiencing a dangerous transformation, through which the global elite has used every tactic available in a conspiracy to hoard an even greater share of wealth and reduce the world's population. And it is working. Many Americans have been drawn not toward life but toward servitude and death. In America and in the world as a whole, entire populations have been culled for profit and control. Elites have used the so-called GOD syndicate—Guns, Oil, and Drugs—as well as toxic air, water, food, and medicines, and of course, the toxic financial system on which the whole master plan depends—to reduce the world's population. This is due to the belief of the global elite that the basis of all the world's problems is overpopulation—just too many people using the earth's limited resources.

Guns, Oil, and Drugs are the top three revenue-generating commodities in the world today, and they form the financial backbone of the global elites. All three are trafficked internationally, generating huge profits for those who control them, and are becoming ever more important in today's economy. America has gone to war for oil, supplied its military (not to mention private citizens) with firearms, and been complicit in a global drug trade. And behind the scenes, a wealthy elite has profited tremendously from all three.

The food, water, and air we consume are also in a state of decay. We believe this decay is normal, the way things have always been. We discount the overwhelming evidence that these are recent developments. We now live in a culture of death and decay that has been imposed upon us by a small group of wealthy elites that publicly espouses involuntary population reduction. We're being killed by chemicals, genetically modified organisms (GMOs), dyes, additives, plastics, tainted water, and polluted air. Numerous common household items are actually powerfully toxic: aspartame, fluoride, GMOs, pesticides, high-fructose corn syrup, pharmaceutical drugs, cell phones, microwaves, even basic electricity. Even though food is plentiful in America, our bodies receive fewer nutrients than they did fifty years earlier.

The commercial products we use every day contain more than eighty-five thousand chemicals; our food and water are poisoned as well.

Until the 1950s, the U.S. was predominately a rural nation. People mostly ate fresh home-grown garden foods. No foods contained genetically modified organisms (GMOs), and chemical food additives were rarely used. Beginning in the late 1940s, corporate food producers increasingly began offering processed, nutrient-deficient foods previously unknown in human history. Children growing up in the 1960s were, for the first time, subjected to imitation foods, processed consumables that appeared and tasted nutritious but lacked any real nourishment. By the 1970s, the American diet consisted of non-nutritious white bread and other unnatural food products. By the 1980s, Americans, raised on imitation foods, drinking fluoridated water, and distracted by disco, had become apathetic and lazy.

We are not aware of these things because precious few recognize that we are being psychologically programmed by a mass media controlled by a mere handful of corporate owners. This handful of multinational media corporations, many with interlocking directors and owners, control everything we see and hear, from movies, TV, and newspapers to satellite networks, magazines, even book clubs and billboards. Indeed, the complicity of the mass media ensures that we cannot protest the population reduction that threatens our very lives.

In fascist Italy and Nazi Germany, the state gained control over the corporations. In modern America, corporations have gained control over the state. The end result is the same.

Giant corporations, owned by a small globalist elite, have thrived often by deceptive and illegal practices. In 1952, corporations accounted for 32 percent of federal tax revenues. By 2013, this number was less than 10 percent. In that same year, forty-six U.S. corporations were blacklisted for corruption by the World Bank. And these corrupt tactics create wealth for only a select few; economic reality in the U.S. today substantiates the old line about the rich getting richer while the poor get poorer.

A 2014 survey by the Russell Sage Foundation found that during the recent economic downturn, lower-income households lost a larger portion of their wealth than those with higher incomes. The study revealed a "startling decline" in wealth nationwide. The median household in 2013 had a net worth of just $56,335—43 percent lower than the median wealth level right before the recession began in 2007, and 36 percent lower than a decade ago. "There are very few signs of significant recovery from the losses in wealth suffered by American families during the Great Recession," concluded the researchers.

Meanwhile, the share of wealth enjoyed by the global elite only increases. What has been described as the "one percent" actually is more like the ".01 percent." What's worse is that one cannot know exactly how much these people are worth, since much of their wealth is hidden in offshore bank accounts. "At the commanding heights of the U.S. economy, hiding a lot of one's wealth offshore is probably the norm, not the exception," noted Paul Krugman in the *New York Times*.

Everyone has heard of some rich Americans: the Rockefellers, Warren Buffett, the Koch brothers, George Soros, and Donald Trump are a few examples of highly visible wealthy Americans. But most of the names of the true one-percenters—the billionaires—are unknown to the public.

Yet these faceless billionaires run the world's financial and political systems, and their wealth and power is only growing greater. The four hundred richest Americans made $200 billion in 2013, a total equal to the combined amount spent on the federal food-stamp, education, and housing programs. Ninety-five percent of all new income generated between 2009 and 2012 went to the wealthiest one percent, who own 38 percent of the nation's financial wealth, while the bottom 60 percent owns just 2.3 percent of the nation's wealth.

Income inequality gained increased public visibility with the 2014 publication of a book entitled *Capital in the Twenty-First Century* by French economist Thomas Piketty, who argued that inequality of capital produces an ever-growing disparity in wealth. This idea does not sit well with those who believe capitalism requires inequality of wealth and that taxes on wealth, capital, inheritance, and property are inimical to growth.

Piketty argues that the response to wealth inequality should be a top income tax rate of up to 80 percent, an effective inheritance tax, increased property taxes, and even a global wealth tax. But he acknowledged such measures are currently inconceivable, as anyone with money wants to keep it, and those in this wealthy elite are the primary financiers of the American political system. It's no wonder that government has done little to curb the power of corporations whose destructive actions are endangering our lives; our political leaders depend upon this corporate blood money for their election.

These corporations, however, are anything but faceless. Every company is owned and operated by individuals, men and women with names and addresses. These persons have family, friends, and private lives. Collectively, they call themselves "globalists," men and women who have a right to dominate based on wealth, heritage, and bloodline. They view the entire planet as their private playing field. They can be identified and located. And they have a plan to control the globe, one formulated many years ago within secretive societies in both Britain and the U.S. It depends upon killing most of us. Here is that plan.

CHAPTER 1

DEPOPULATION

THE GEORGIA GUIDESTONES IS A MONUMENT IN ELBERT COUNTY, Georgia. It is composed of four sixteen-foot-tall stones that have been called the American Stonehenge. Indeed, its origin is as mysterious as its English namesake. Commissioned in 1979 by a man using the pseudonym R. C. Christian, the monument was constructed by the Eberton Granite Finishing Company and completed in 1980. An accompanying tablet states that the sponsors of the stones are "a small group of Americans who seek the Age of Reason." A message is inscribed on the stones in eight modern languages and four ancient ones.

Below the title *Let These Be Guidestones to the Age of Reason*, the engraved message reads:

MAINTAIN HUMANITY UNDER 500,000,000 IN PERPETUAL BALANCE WITH NATURE.

GUIDE REPRODUCTION WISELY—IMPROVING FITNESS AND DIVERSITY.

UNITE HUMANITY WITH A LIVING NEW LANGUAGE.

RULE PASSION—FAITH—TRADITION—AND ALL THINGS WITH TEMPERED REASON.

PROTECT PEOPLE AND NATIONS WITH FAIR LAWS AND JUST COURTS.

LET ALL NATIONS RULE INTERNALLY RESOLVING EXTERNAL DISPUTES IN A WORLD COURT.

AVOID PETTY LAWS AND USELESS OFFICIALS.

BALANCE PERSONAL RIGHTS WITH SOCIAL DUTIES.

PRIZE TRUTH—BEAUTY—LOVE—SEEKING HARMONY WITH THE INFINITE.

BE NOT A CANCER ON THE EARTH—LEAVE ROOM FOR NATURE—LEAVE ROOM FOR NATURE.

Some view the stones as offering reasonable and rational suggestions for developing a peaceful and just world. Others see something more sinister. One conspiracy website noted astronomical features within the stones. The four major stones are oriented to reflect the migration limits of the sun during the year, while a hole in the center stone always aligns with the North Star and another hole aligns with the rising sun during the summer and winter solstices. Such celestial alignments are found in the works of secret societies from the Freemasons to the Druids and the Mystery Schools of ancient Greece and Egypt.

"The monument is therefore proof of an existing link between secret societies, the world elite and the push for a New World Order," declared the website. In 2008, vandals defaced the monument with the words "Death to the New World Order."

The stones' first admonition is the most disconcerting to many, as the world population in mid-2014 stood at nearly seven and a half billion persons. If the Guidestones' mandate to hold the human population to five hundred million is achieved, what is to happen to the other seven billion?

England's Prince Philip, the Duke of Edinburgh and a prominent globalist, may have revealed the views of the global elite when in 1981 he told *People* magazine, "Human population growth is probably the single most serious long-term threat to survival. We're in for a major disaster if it isn't curbed—not just for the natural world, but for the human world. The more people there are, the more resources they'll consume, the more pollution they'll create, the more fighting they will do. We have no option. If it isn't controlled voluntarily, it will be controlled involuntarily by an increase in disease, starvation and war."

Years later, Philip mused, "In the event that I am reincarnated, I would like to return as a deadly virus, in order to contribute something to solve overpopulation."

As one of the founders of the World Wildlife Fund (WWF), Prince Philip once laid out a globalist justification for depopulation. "The object of the WWF," he wrote, "is to 'conserve' the system as a whole; not to prevent the killing of individual animals. Those who are concerned about the conservation of nature accept . . . that most species produce a surplus that is capable of being culled without in any way threatening the survival of the species as a whole."

Other globalist leaders agree with this chilling assessment. In a 1981 interview concerning overpopulation, former chairman of the Joint Chiefs of Staff Maxwell Taylor said by the beginning of the twenty-first century, it would be necessary to reduce the world's population, mostly in third-world countries, using methods such as disease, starvation, and regional wars. He blithely concluded, "I have already written off more than a billion people. These people are in places in Africa, Asia, and Latin America. We can't save them. The population crisis and the food-supply question dictate that we should not even try. It's a waste of time."

Such brutal tactics have even been incorporated into national policy in some countries, including the United States. In 1974, the U.S. National Security Council issued a classified study entitled "National Security Study Memorandum (NSSM) 200: Implications of Worldwide Population Growth for U.S. Security and Overseas Interests." Known as the Kissinger Report, the study stated that population growth in the so-called Lesser Developed Countries (LDCs) represented a serious threat to U.S. national security. The study was adopted as official government policy in November 1975 by President Gerald Ford and its implementation assigned to Brent Scowcroft, who had replaced Kissinger as national security adviser. NSSM 200 outlined a covert plan to reduce population growth in LDCs through birth control, and what many have interpreted as war and famine. Then CIA director George H. W. Bush was ordered to assist Scowcroft, as were the secretaries of state, treasury, defense, and agriculture. This policy may even have supported the many wars and airstrikes in the Middle East leading to a decimation of the populations there.

There is even significant evidence that claims of overpopulation are spurious. It seems the real issue is one of population density rather than population growth. For example, according to the Oklahoma Department of Agriculture, the state covers an area of 69,903 square miles. If each person is allowed one hundred square feet of living space, Oklahoma could accommodate 19.49 billion people—nearly three times the earth's current population of seven billion.

Of course, this merely illustrates that that the earth still has plenty of room for everyone, not that everyone would want to live in one state. If the world's population could spread out, and avoid concentrating in sprawling metropolitan centers, citizens would most likely be much happier and better off. As is, crowding in cities produces the unwelcome effects of crime, congestion, pollution, and stress. Studies have shown that lab rats are content in their cages until too many in too close contact cause them to turn on each other.

Yet, leading one-percenters continue to echo the tone of NSSM 200 and Prince Philip's remarks. On May 5, 2009, some of America's leading billionaires met in a private Manhattan home just a week before the annual meeting of the secretive Bilderbergers. Calling themselves the "Good Club," attendees reportedly included Bill Gates, David Rockefeller Jr., Warren Buffett, George Soros, New York mayor Michael Bloomberg, Ted Turner, and Oprah Winfrey. According to John Harlow of the *Sunday Times,* the group—while not going so far as to advocate active depopulation strategies—agreed with Gates that human overpopulation was a priority concern. Harlow said there was nothing as crude as a vote but a consensus was reached that "they would back a strategy in which population growth would be tackled as a potentially disastrous environmental, social and industrial threat."

Apparently, those with great wealth and power have decided to take overpopulation into their own hands. Dave Hodges, host of The Common Sense Show, recalled how President Ronald Reagan once remarked that a threat by space aliens might be the unifying force necessary to bring the nations of the Earth together in a common cause. Hodges warned, "Indeed, all of mankind does face a common foe. However, it is not aliens. Our common foe is the elite that presume that they have the God-given right to exert ownership over all of us including the right to life or death. And for 95 percent of us, the elite are actively engaged in systematic extermination of mankind."

And some of today's elite can be connected to the same families and corporations that funded communism in Russia and then national socialism in prewar Germany.

In noting the similarities between the rise of the Nazis and modern America, Dr. Len Horowitz said, "Today with AIDS, mad cow disease, chronic fatigue, and the rest, history is apparently repeating. In fact, even the message is the same. The millions of Holocaust victims were told they were going into 'showers' for 'public health' and 'disinfection.' That's why we are being told to get vaccinated. Virtually nothing has changed, not even the message."

The late Donald W. Scott, schoolteacher and author of The Brucellosis Triangle and a Canadian political candidate, has speculated that as far back as the 1940s there existed a high-level agenda to research a viral pandemic of brucellosis by testing it on unwitting U.S. citizens, a project Scott suggests was initiated by persons holding sway over government officials: "The Washington corner of the brucellosis triangle with its military, NIH [National Institutes of Health], Treasury and Justice [Department] components have had their ties to and have largely taken their directions from the New York corner dominated by the Rockefeller interests. And the Rockefeller interests through the agency of the CFR [Council on Foreign Relations], the Rockefeller Institute/University, the Cold Spring

Harbor Laboratory, the Rockefeller Foundation and the Chase Manhattan Bank [now simply Chase] have constituted a vast machine of power and baleful influence whose parts have meshed together in an effort to maintain that power."

As detailed in my book *The Rise of the Fourth Reich,* the Rockefeller family laid the foundation for many of America's major medical institutions beginning at the end of the Civil War. Besides funding universities and the eugenics movement, Rockefeller largess includes such entities as the Rockefeller Sanitary Commission, the Rockefeller Institute for Medical Research (now Rockefeller University), and the General Education Board, which expended massive funds on medical schools to produce doctors inclined to allopathic medicine (the predominant use of drugs and surgery).

But many question whether overpopulation truly is a problem of the magnitude being argued by the wealthy one percent and their corporate mass media.

In mid-2014, *Business Insider* published an article by Marian Swain, a conservation and development policy analyst for the Breakthrough Institute, a think tank dedicated to modernizing environmentalism for the twenty-first century.

Swain reported that while the world population continues to grow, the *rate* of growth actually has been decreasing since a peak in the 1960s. Between 1965 and 1970, the world population growth rate increased by 2.1 percent. Currently, the world population is growing at half this number, only about 1.2 percent per year. She wrote, "We are already experiencing a slowdown in population growth, and it is expected to continue in the coming decades. The UN's median scenario projects flat or decreasing population size in all regions except Africa. Other projections suggest that the global population may even peak this century."

She also noted that new technology may increase the earth's food-producing capacity, alleviating fears that food production methods will be unable to keep pace with population growth. "It is sometimes suggested that there are hard biological limits to how much food the earth can produce, but ever since the invention of agriculture 10,000 years ago humans have been consistently increasing yields through the use of new technologies, such as herbicides, growth stimulants and mechanization. Indeed, it has been increasing yields that have allowed the human population to grow to its current population of seven billion. In this sense, the earth's carrying capacity is not bound by a finite set of planetary boundaries, but rather is a function of human technology," she wrote.

She added that while population is undoubtedly a factor in anthropogenic climate change, as human activities do create greenhouse-gas emissions, a far larger factor is the kind of energy being used. "One billion people on the planet getting electricity from coal would create more carbon emissions than 6 billion people each getting the same amount of electricity from solar or nuclear power. To combat climate change, technology is more important than population."

Swain also noticed that fertility (the average number of children a woman gives birth to in her life) is closely correlated with development. She said UN statistics show "the countries with the highest fertility rates are generally the poorest ones, while almost all the richest countries have fertility rates that are actually below the replacement rate of 2.1 children per woman."

Citing a clear correlation between fertility and development, Swain notes that as incomes around the world increased between 1910 and 2010, fertility rates fell dramatically. "In the developing world, people are increasingly moving to cities, gaining access to modern services, and the fertility rates of these countries have, in turn, been falling . . . There is even evidence that exposure to modern media like television can create downward pressure on family size," she wrote, adding, "This is not to say that we should not do anything to promote lower birth rates . . . However, access to contraception is only one of many factors that affect women's fertility choices. Broader issues of poverty and education are also crucial to address if we hope to encourage women to choose smaller family sizes."

Yet despite evidence that fears of population growth are overblown, the globalists seeking population reduction have continued their systematic elimination of huge numbers of people. This population reduction has taken many forms. Following the 2009 outbreak of swine flu (H1N1 influenza virus), it was found that the strain contained a combination of genes from swine, bird, and human influenza viruses. Because this virus could not be contracted by eating pork or pork products, researchers suspected swine flu was manufactured by humans. They believed the outbreak was one of several venues being used to reduce the human population by the global elite, who have long supported eugenics, the social philosophy of improving genetic traits by eliminating less desirable people.

In the early 1970s, Associate Supreme Court Justice Ruth Bader Ginsburg may have betrayed the views of many globalist intellectuals when she said she believed the *Roe v. Wade* abortion decision was predicated on the Supreme Court majority's desire to diminish "populations that we don't want to have too many of." She added that it was then her expectation that the right to abortion would later be expanded to "Medicaid funding for abortion."

Where did Ginsburg get the idea that American policy-making elites were interested in decreasing undesirable populations? Some researchers suggest that Ginsburg, at some point, became acquainted with the writings of John Holdren or other similar writers in the most militant branch of the population control movement. In 1977, Mr. Holdren was a young academic who assisted birth control guru Paul Ehrlich and his wife Anne in writing *Ecoscience: Population, Resources, Environment.*

In this book, Ehrlich wrote, "Many of my colleagues feel that some sort of compulsory birth regulation would be necessary to achieve such control [over population growth]. One plan often mentioned involves the addition of temporary sterilants to water supplies or staple food. Doses of the antidote would be carefully rationed by the government to produce the desired population size." Expressing the desire for "a Planetary regime" by controlling all human economic activity and interactions with the environment, the Ehrlichs and Holdren urged governments to use "power to enforce the agreed limits" on population growth by whatever means necessary, including involuntary sterilization, abortion, or even mass involuntary sterilization through chemicals in public water supplies.

With Holdren contributing, the book noted "a program of sterilizing women after their second or third child . . . might be easier to implement than trying to sterilize men" and that "compulsory population-control laws, even including laws requiring compulsory abortion, could be sustained under the existing Constitution if the population crisis became sufficiently severe to endanger the society." In 2009, during Senate confirmation hearings, Holdren tersely claimed he had renounced such views.

It should be pointed out that amid the Obama administration's efforts to impose centralized and universal Obamacare, John Holdren sits as the president's director of the White House Office of Science and Technology Policy. As "science czar," Holdren counsels the president on the role of science in public policy. "This relationship has a certain Strangelovian undercurrent, given Holdren's enthusiasm for eugenicist and totalitarian methods of population 'management,'" notes Internet blogger and radio host William Norman Grigg.

G. Edward Griffin, author of *The Creature from Jekyll Island*, a history of the Federal Reserve System, has also voiced concern over Holdren's thoughts on martial law and depopulation. Noting Holdren's early mention of forced abortions and putting sterilization chemicals in the water supply in the 1970s book, Griffin stated that Holdren seemed to have advanced the idea of reducing the population by insidious means. "He was not concerned with the ethical or freedom issues involved with these measures, only their practicality. Now we find this same man, an academic expert on population reduction, at the right hand of the President of The United States," Griffin notes.

And he adds, "Remember, all of those who hold power in the governments of the world today [the self-styled globalists] are collectivists and the guiding rule of collectivism is that individuals and minorities must be sacrificed, if necessary, for the greater good of the state or of society. Of course, those who rule will decide what the greater good is and who is to be sacrificed."

This, of course, is the basic problem with population control. The idea of limiting the burgeoning earth's population may appear desirable, as the increasing number of humans as well as their waste continues placing a strain on the planet. The burning question is *who* gets to decide which segments of the population must forgo childbearing for the good of the majority. So far, it is the wealthy globalists who have taken the lead in supporting ways to hold down population growth through eugenics, drugs, and birth-control measures.

Catherine Austin Fitts, who served as former Assistant Secretary of Housing under the administration of George H. W. Bush, has explained why depopulation may be one of the globalists' foremost goals. "My simple calculations guessed that we were going to achieve economic sustainability on Earth by depopulating down to a population of approximately 500 million people . . . I was . . . used to looking at numbers from a very high level. To me, we had to have radical change in how we governed resources or depopulate. It was a mathematical result."

Fitts noted that some government budget analysts have concluded that the nation can no longer afford social safety nets like Social Security and Medicare. "That is, unless you change the actuarial assumptions in the budget—like life expectancy," she said. "Lowering immune systems and increasing toxicity levels combined with poor food, water and terrorizing stress will help do the trick."

She envisioned a pandemic that would so frighten the public that they could be controlled and even accept the end of current government benefits. The growing scares over Ebola, E. coli, and the various deadly influenzas may eventually achieve this end.

Some very strange and scary incidents that do not portend well for population growth have been reported. For example, in 2012, Sara Barron, then a thirty-year nursing veteran, witnessed separate incidences of anencephaly, a horrible birth defect in which babies are born missing parts of their brain and skull. Barron had encountered this problem only twice in her career. Now there were two cases within two months and in the same small rural hospital in Washington State.

Other such cases were found and the state department of health was notified. They discovered that between January 2010 and January 2013, there were 23 cases of anencephaly in a three-county area.

This meant a rate of 8.4 cases per ten thousand live births—four times higher than the national average.

The puzzle deepened when no cause could be immediately discovered for this tragic condition. Furthermore, there was criticism that not enough was being done to locate a cause. Mothers of these babies said they were never contacted by medical authorities investigating the brainless births.

A study of medical records proved fruitless, which did not surprise Dr. Beate Ritz, vice chair of the epidemiology department at the UCLA Fielding School of Public Health. Ritz said such records are notoriously unreliable. She added that state health departments simply do not have the money to conduct in-depth research.

Meanwhile, Nurse Barron said, "I think it's very scary. I think there's absolutely something going on that needs to be investigated more thoroughly. I wish they would take it more seriously."

In addition to strange diseases and conditions, deadly food additives, contaminated water, and vaccines all are contributing to population reduction, as will be described in the following pages. One must ask if this is simply coincidence or inattention, or if there is a conscious agenda to depopulate the world.

CHAPTER 2

THE DEADLY GOD SYNDICATE

THE WORLD SPENDS THE INCOMPREHENSIBLE SUM OF ABOUT $1 trillion annually on military hardware. This includes small arms, armored vehicles, ships and submarines, and aircraft. Arms procurement can represent up to 30 percent of a nation's military budget.

The United States has been the world's largest arms dealer for many years, peddling more weapons than Russia and China combined. Between 2003 and 2011, the United States ranked first in arms transfer agreements with developing nations, with U.S. agreements over this period worth a total of $56.3 billion, or 78.7 percent of the value of all such agreements worldwide. And America's closest competitors over this time frame were not really competitors at all. Russia ranked second with $4.1 billion in arms transfers, or just 5.7 percent of such agreements. China, often said to be a threat to the U.S., registered only a measly 3 percent.

Due to the current global economic downturn, many weapons-exporting nations, facing increased competition, have begun expanding into new markets. Richard F. Grimmett, author of a Congressional Research Service report on the matter, noted that despite a global decline in arms sales in 2011, the U.S. recorded an "extraordinary" increase in market share, primarily due to massive sales to Saudi Arabia and India. Such increased arms sales indicated an effort to exert American influence in both the Middle East and in India, the largest block to Chinese expansion in the East.

The numbers above account only for government-to-government foreign military sales (FMS). These statistics do not include private or illegal sales, which are substantial, and came to public attention during the "Fast and Furious" gun-walking scandal in 2012, in which the U.S. government was complicit in allowing guns to pass into the hands of Mexican drug cartels. According to Transparency International, an organization that monitors corruption, the international trade in armaments is among the most corrupt businesses in the world. Illegal arms transfers undermine many developing countries' chances of achieving their development goals by draining their resources, and in some cases, fueling armed conflict.

BLACK MARKET ARMS

THE U.S. HAS THE DUBIOUS HONOR OF BEING THE LEADER NOT ONLY in legal arms sales but also in the shadowy world of black-market weapon sales. These illegal arms often fall into the hands of America's worst enemies, including terrorists.

Recent international events have underscored U.S. involvement in the illegal international arms trade. The 2011 ouster of Libyan leader Muammar Gaddafi, the 2012 murder of U.S. ambassador Christopher Stevens in Benghazi, and the Obama administration's arming of Syrian rebels attacking the government of Syrian president Bashar al-Assad were all connected to under-the-table transfers of arms by the United States.

Various sources allege that a program known as Direct Commercial Sales (DCS) is behind this bloody turmoil. This group operates within the U.S. State Department's Directorate of Defense Trade Controls (DDTC). The DCS program regulates private U.S. companies' overseas sales of weapons and other defense articles, defense services, and military training. It is separate from the Foreign Military Sales (FMS) program, which manages government-to-government sales. Through DCS, vast sums of money are shuffled through international banks, multinational corporations, and foreign governments.

According to a report by the American Federation of Scientists, the State Department is much less transparent about DCS than the Pentagon is about FMS. "Minimal information about price and quantity is classified as 'confidential business information' and kept from the public. This secrecy undermines the ability of Congress and the interested press and public to exercise proper oversight on industry-direct arms transfers."

In mid-June 2013, the White House announced that President Obama had authorized "direct military support" to Syrian rebel forces, thus allowing DCS to operate in that Middle East nation. According to a Reuters news dispatch, "Syrian rebel and political opposition leaders immediately called for anti-aircraft and other sophisticated weaponry. The arrival of thousands of seasoned, Iran-backed Hezbollah Shi'ite fighters to help Assad combat the mainly Sunni rebellion has shifted momentum in the two-year-old war, which the United Nations said . . . had killed at least 93,000 people."

On September 17, 2013, the White House announced that President Obama had waived portions of a federal law aimed at preventing the sale of arms to terrorist groups. He did this so that the Syrian rebels could legally be supplied weaponry and ammunition. This waiver could prove problematic, according to the *Washington Examiner,* since a significant portion of the Syrian opposition has been connected to radical Islamic terrorist groups including al-Qaeda and ISIS [the Islamic State of Iraq and al-Sham].

For several years American intelligence agents operating from a number of safe houses in Syria aided in the sale of arms to the Syrian rebels even to the extent of deciding which terrorist gang or commander should receive the weapons as they arrived. The *New York Times* in March 2013 reported the scale of arms shipments was "very large," and that the Turkish government exercised oversight over much of the operation. "A conservative estimate of the payload of these flights would be 3,500 tons of military equipment," stated Hugh Griffiths, an illicit arms transfers monitor for the Stockholm

International Peace Research Institute. "The intensity and frequency of these flights are suggestive of a well-planned and coordinated clandestine military logistics operation."

In early 2015, the Citizens Commission on Benghazi (CCB), a group of private citizens that included former military commanders and Special Forces operatives, former CIA and intelligence officers, international terrorism experts, and persons knowledgeable in media and government affairs, confirmed that U.S. officials were providing weaponry to American's enemies. A CCB interim report entitled "Changing Sides in the War on Terror" concluded that the Obama White House and the State Department under the management of Secretary of State Hillary Clinton "changed sides in the war on terror" in 2011 with a policy of sending weapons to the al-Qaeda-dominated rebel militias in Libya attempting to oust Muammar Gaddafi from power.

"The rebels made no secret of their Al Qaeda affiliation," said report author John Rosenthal. "And yet, the White House and senior congressional members deliberately and knowingly pursued a policy that provided material support to terrorist organizations in order to topple a ruler who had been working closely with the West actively to suppress al Qaeda." Some claim Gaddafi was overthrown with U.S. assistance because he was about to create an African "dinar" backed by gold that would have undercut the U.S. dollar.

"Stevens was facilitating the delivery of weapons to the al-Qaida-related militia in Libya," confirmed Clare Lopez, a former CIA operations officer and member of the commission who is currently vice president for research at the Washington-based Center for Security Policy.

Kevin Shipp, a former CIA counterintelligence expert, and Lopez both agreed that the gunrunning operation bordered on treasonous activity and is a secret the Obama White House and Clinton State Department sought to suppress from the public.

In the "Blue Lantern" program, the DDTC monitors end-use recipients of weapons and services licensed by the State Department and provided by DCS. This program is intended to ensure that arms do not fall into the wrong hands. But some defense industry sources now claim that DCS is "playing both sides against the middle for corporate or political gain."

William Robert "Tosh" Plumlee, a former CIA contract pilot who flew arms and ammunition for the agency as far back as the overthrow of Cuban dictator Fulgencio Batista and the 1961 Bay of Pigs invasion, questioned if such arms dealing might be another "off-the-books" covert operation run by the CIA's Special Tactical Unit akin to the arms-for-drugs deal in Iran-Contra and the Cuban Project of the 1950s, in which both Fidel Castro and the Batista government in Cuba were sold weapons from American stockpiles for corporate profit. Numerous field reports have stated Plumlee flew arms to Nicaragua during the Iran-Contra Scandal. In testimony to the U.S. Senate, Plumlee also said he returned to the U.S. with loads of cocaine during the Reagan years. Recently, Plumlee has worked as a photojournalist along the U.S.-Mexican border and participated in investigations into the Bureau of Alcohol, Tobacco, Firearms and Explosives' (ATF) Fast and Furious sting operation.

Blue Lantern reports, which date back to the early 2000s, confirm that many investigations of the end users of weaponry supplied by U.S. firms were "unfavorable," with arms sometimes ending up in the hands of foreign enemies. Although law enforcement agencies receive these reports in order to evaluate possible legal action, investigations are usually dropped due to foreign-relations considerations.

Even lawmakers, who supposedly work for the public good, are involved in the arms trade. Many legislators own stock in armaments firms. Some are more intimately involved, such as California state senator Leland Yee, who in 2014 was indicted by a San Francisco grand jury for corruption and conspiracy to traffic in firearms. The irony of Yee's plight was that the Democratic politician was an advocate of stricter gun control.

Yet a rampant and corrupt American trade in arms is the least of our problems.

Despite George Washington's parting advice to beware foreign entanglements, the United States since World War II has followed a foreign policy of interventionism and adventurism that has only benefited the arms manufacturers. As of 2011, the U.S. had active military troops stationed in nearly 150 nations, including small countries such as Albania, Croatia, Estonia, and Ireland.

Perpetual war allows globalists to continue funding dirty black-ops drug smuggling, corrupt banking practices, political bribes, and assassinations. Perpetual war can be seen as an excuse for spying on Americans, militarizing police agencies, and laws allowing the federal government to declare any American citizen an "enemy combatant" and holding them without warrant or habeas corpus as well as spying with drones.

With secretive societies, such as the Council on Foreign Relations, providing leadership for both the Democratic and Republican parties, there has been no significant change in U.S. foreign police since World War II. The global elite that control both parties sees to it that no one who is not aligned with globalist goals gains the presidency. No effort is spared to keep America in perpetual war, the basis for the elite's global agenda.

Investment in infrastructure would be a far better use of federal funds than investment in the military. The nation's highways, dams, and bridges continue to deteriorate, with many receiving failing grades from the American Society of Civil Engineers (ASCE).

In 2013, the ASCE, committed to protecting the health, safety, and welfare of the public by improving the nation's public infrastructure, issued its "report card" grade based on physical condition and needed investments for improvement. The USA got a D-plus.

Yet the proposed 2015 defense budget is more than $600 billion and protects a long list of weapons programs. This budget also includes such items as $69 million for a new prison facility at Guantánamo Bay, Cuba, designed to house a mere fifteen "high-value" prisoners, and a $2 billion NSA data center at Bluffdale, Utah, to store Americans' intercepted email, text, and phone messages. With budgetary decisions such as these, it is apparent that the U.S. government values its position at the forefront of military technology more than it values the lives of its citizens.

PRIVATE GUN OWNERSHIP

AMERICA'S INFATUATION WITH WEAPONRY IS PERHAPS BEST EXEMPLI-
fied by how many private citizens own guns. The U.S., despite having
less than 5 percent of the world's population, has roughly 35 to 50
percent of the world's civilian-owned guns. Yet it's not at all clear
from the global statistics that private gun ownership can be equated
with violence. The countries with the third and fourth highest rates
of gun ownership may be unexpected: Switzerland and Finland,
which have some of the lowest crime rates in the world. A similar
link between gun ownership and *reduced* crime can be found in FBI
statistics, which showed only one gun-related homicide during 2012
in Alabama, a state lenient on firearms, versus 1,304 such deaths in
California, a state with some of the strictest gun laws.

Chicago is another prime example of the ineffectiveness of gun
control laws. Despite some of the most stringent antigun laws in the
nation, Chicago led the nation in shootings in the first six months
of 2014, with more than 1,100. During the July 4, 2014, weekend
alone, there were 84 shootings and 14 homicides in Chicago. Yet
the corporate mass media failed to inform the public that Chicago,
with some of the strictest gun control laws in the country, routinely
has more shooting deaths than other cities that recognize a citizen's
natural right of self-defense by allowing them to freely and openly
carry a personal defense weapon.

In recent years, school shootings have provided another talking
point for both sides of the gun control debate. Those in favor of more
stringent gun control cite the spate of recent shootings as evidence of
our need for stricter restrictions. Meanwhile, pro-gun groups argue
that such shootings would decrease if would-be shooters knew that
every school contained a teacher, coach, or principal who was trained
and armed.

In years past, many students, particularly in the south and west,
carried guns to school, most often in the racks in pickup trucks for
after-school hunting. According to former attorney general Eric
Holder, the yearly average of mass shootings in the U.S. tripled in

recent years, from an average of five per year between the years of 2000 and 2008 to twelve mass shootings just in 2013.

Researchers at Harvard University in October 2014 reported that mass shooting incidents have increased threefold since 2011. They said on average a mass shooting took place every sixty-four days during this period, compared with an average of every two hundred days in the years from 1982 to 2011.

As will be demonstrated later, the true cause of the recent rise in mass shootings is not weapons but the increase in psychiatric drugs being prescribed for youngsters.

The effectiveness of guns as a deterrent to crime has been proven in Kennesaw, Georgia, which in 1982 passed an ordinance requiring heads of households (with some exceptions) to keep at least one firearm in their homes. By 2001, violent crime rates in Kennesaw had dropped to about 85 percent below national and state rates while property crime dropped to about 50 percent below national and state rates. This decrease generally continued through 2012, with the exception of some slight increase between 2003 and 2008, accounted for by population growth twice the national average. Though there are numerous other stories like the one in Kennesaw, the globalist-controlled mass media, with its antigun agenda, almost never reports them.

In early 2013, thirty-three-year-old Deyfon Pipkin, who had a lengthy criminal record, was killed with a single shot by an elderly homeowner in Dallas after breaking into the man's home. Pipkin's family bemoaned the lack of a warning shot. "He could have used a warning," Pipkin's sister-in-law, Lakesha Thompson, complained to the media. "He could have let him know that he did have a gun on his property and he would use it in self-defense." Others wondered why Pipkin's family had not warned him about the consequences of breaking into people's homes to commit crimes.

In April 2014, forty-year-old Mitchell Large, a man whom authorities said had a lengthy criminal record for domestic violence and assault, was fatally shot by members of the Luis Peña family after he broke into their Winter Haven, Florida, home. Police Chief Gary Hester said the father, mother, and adult son all armed themselves, and a warning shot was fired, but the intruder continued into the house. No charges were filed in the Peña case, and Hester told the media: "Whether [Large] was armed or not armed, when he failed to retreat they certainly had a right . . . to defend themselves."

In America, the mere presence of firearms does not equate to increased homicide rates. But then statistics, reason, and common sense do not seem to apply when it comes to the debate on ever-increasing gun control, a favored globalist agenda reaching all the way to the United Nations. In September 2014, Secretary of State John Kerry signed the long-delayed UN Arms Trade Treaty, intended to curb the international sales of weapons, prompting the National Rifle Association's Chris Cox to declare the treaty a "global gun grab treaty" and a "blatant attack on the constitutional rights and liberties of every law-abiding American."

Lastly, the two principal reasons behind most gun violence— stress due to poverty intensified by alcohol consumption—are largely ignored by the corporate mass media. This is because movements to address poverty do not pay for advertising, unlike the alcohol industry. Poverty and the unequal distribution of wealth create stress on even the most functional of families, especially in cash-strapped cities such as Detroit, Chicago, and Minneapolis. The poorer sections of major cities also experience more gun violence due to stressful living conditions.

According to Robert Nash Parker in a paper entitled "The Effects of Context on Alcohol and Violence," published in a 1993 issue of *Alcohol Health & Research World,* "Alcohol consumption increases violence within the context of poverty, and violent behavior may be perceived as a rational and acceptable choice in some contexts."

Yet the media never reports on these true causes of gun violence. It is clear that the self-styled "globalists" who own and control the corporate mass media have a duplicitous agenda. While they profit from the international trade in arms, domestically their policy is very different: they seek to demonize guns to precipitate a cry for more stringent laws and gun registration, to be followed by confiscation. After all, a disarmed population is more easily controlled.

CHANGING THE GAME

IT IS CLEAR THAT THE PROBLEM OF GUN VIOLENCE HAS LESS TO DO with the availability of weapons than with the hopelessness of poverty and the polarization of wealth. Dorothy Stoneman is the founder of YouthBuild USA, a program in forty-six states offering low-income young people jobs while they work toward a high school diploma. She explained, "If America spent as much money offering opportunities to every sixteen- to twenty-six-year-old as we spend locking them up for minor offenses that further cut them off from a positive future, we could end poverty in a generation or two. When young people find a true pathway to opportunity and a caring community, they become excellent parents determined to give their children the world of opportunities they lacked in their own childhood."

It is true that guns do not kill people; people kill people. Until we confront and resolve sources of societal discontent, no antigun legislation will keep the public safe from lawless gun-toting criminals or the mentally unbalanced. Yet neither the corporate media nor self-serving politicians are interested in addressing the basic problem of poverty. Politicians have always found it easier to simply pass more laws instead of probing the true causes of gun violence.

No one should expect government to end arms production. Progress toward a more peaceful and nonviolent society must begin at the local level. Alternatives should be found for arms dealers. The manufacturers of war materials could shift to producing more socially beneficial products. Yet even if the arms trade can be brought under control, we will still face issues related to the second part of the GOD syndicate, and the second most profitable commodity in the world—oil.

OIL

FOR MANY YEARS, SOME PETROLEUM EXPERTS HAVE CLAIMED THE world's supply of oil has peaked and is now in decline.

A decline in oil availability would lead to higher energy prices and worldwide instability. Should the supply of petrochemicals decline, new energy sources would need to emerge to fill the gap. We've reached a point where nations are addicted to oil, in part because the most profitable business on the planet is arms, and all war machines run on petroleum, either as fuel or lubricants.

But war machines are only a small part of the picture, as petroleum provides the foundation for modern civilization. Computers and TVs are made from it, as are all plastics, food wrapping, shampoo, garbage bags, clothes softeners, some furniture, most medicines, and even water bottles. These household necessities reach us by traveling the nation's roadways via trucks, which are also fueled by petroleum. Our economic system is so heavily dependent upon oil that if there were to be a shortage, the price of virtually every good would rise. And petroleum consumption will only increase as the world's population continues to rise.

PEAK OIL

THE TERM "PEAK OIL" WAS COINED BY AMERICAN GEOPHYSICIST Marion King Hubbert, who in 1956 predicted a peak in U.S. oil production by 1970 followed by steady decline worldwide. Initially, many petroleum experts scoffed at the Hubbert Peak Theory, but today it is more respected, even though his specific projection has proven false. While the year 2005, in which global production of oil indeed declined seventy-four million barrels per day, was cause for alarm, production has since recovered, setting new records in both 2011 and 2012.

Some experts claim the only spare oil production capacity left in the world is in the Organization of the Petroleum Exporting Countries (OPEC), composed primarily of Middle Eastern nations. Peak oil advocates believe that non-OPEC oil production limits have already been reached.

"All the easy oil and gas in the world has pretty much been found. Now comes the harder work in finding and producing oil from more challenging environments and work areas," said William J. Cummings, a spokesman for ExxonMobil.

Fear over peak oil has benefited those in the oil industry, who can consequently charge higher prices and justify further exploration for petroleum sources. But the foundations of the concept are weak. Cambridge Energy Research Associates (CERA), composed of energy expert consultants, has been critical of the theory. "Despite his valuable contribution, M. King Hubbert's methodology falls down because it does not consider likely resource growth, application of new technology, basic commercial factors, or the impact of geopolitics on production. His approach does not work in all cases—including on the United States itself—and cannot reliably model a global production outlook. Put more simply, the case for the imminent peak is flawed. As it is, production in 2005 in the Lower 48 in the United States was 66 percent higher than Hubbert projected," stated a 2006 CERA report. The International Energy Association (IEA) also doubts the existence of peak oil, arguing that the global production has reached a plateau rather than a peak.

Hubbert's theory has allowed oil companies to profit from stoking public fear. Yet the remarkably effective new techniques being used to extract oil prove that we have nothing to worry about.

THE BAKKEN FORMATION

THE RECENT DISCOVERY OF THE BAKKEN ROCK FORMATION, WHICH underlies North Dakota, as well as parts of Montana, Saskatchewan, and Manitoba, has made North Dakota the number two oil-producing state in the U.S., behind only Texas. The Bakken is the largest domestic oil discovery since Alaska's Prudhoe Bay, and has the potential to eliminate all American dependence on foreign oil. The Energy Information Administration (EIA) estimates its potential at 503 billion barrels. Just 10 percent of this oil would be 50 billion barrels. If sold for $107 a barrel, this would mean a resource worth more than $5.3 trillion. This one field could provide enough energy to last more than two thousand years even with increased consumption.

In view of this new oil source, there is increasing evidence that petroleum reserves are much greater than noted by the corporate mass media. The political ramifications of this are staggering. After all, it is the specter of worldwide oil shortages that fuels the arguments of New World Order stalwarts who have argued for both centralized government and private programs in the name of conservation and environmentalism.

One example is former presidential candidate Al Gore, who has been an advocate of global warming as well as the Chicago Climate Exchange (CCX), styled as "North America's only voluntary, legally binding greenhouse-gas reduction and trading system."

This attempt to capitalize on carbon restrictions ended in November of 2010 when CCX shut down its operations following the failure to win cap-and-trade legislation in the Republican-controlled House of Representatives.

Despite the overabundance of oil, as exemplified by the Bakken formation, the oil industry continues to profit from fears over peak oil, while the search for new energy resources continues with deadly effects.

FRACKING

AN EXTRACTION TECHNIQUE THAT HAS GAINED CURRENCY IN recent years is fracking. The term, short for "hydraulic fracturing," is the process of drilling and injecting fluid (usually water along with chemicals) into the ground at a high pressure to fracture shale rocks, thereby releasing the natural gas inside. Each fracking site requires between one and eight million gallons of water, mixed with sand and about forty thousand chemicals, including toxic substances such as lead, uranium, mercury, ethylene glycol, radium, methanol, hydrochloric acid, and formaldehyde. More than half of these chemicals are nonbiodegradable and remain in the ground. Recovered fracking waste is left in open-air pits to evaporate. This releases volatile organic compounds, creating contaminated air, acid rain, and ground-level ozone. Metal ball bearings are added to fracking fluids to keep the fractures open indefinitely.

These toxic chemicals leach out and contaminate nearby groundwater. Drinking-water wells near fracturing sites have been found to contain concentrations of methane seventeen times higher than normal.

About 80 percent of the 155,000 wells in the United States involve the use of wastewater to recover hydrocarbons through hydro-fracking, a technique initially developed by Halliburton. Significant evidence exists that corporate owners are using fracking to destroy water wells and sicken the population in an effort to sell more pharmaceuticals while cornering the market on bottled water.

More than a thousand incidents of water-well contamination have been documented near fracking wells along with reports of sensory, respiratory, and neurological damage due to those drinking the water. According to the Natural Resources Defense Council, residents in Arkansas, Colorado, Pennsylvania, Texas, Virginia, West Virginia, and Wyoming have reported health problems along with negative changes in water quality and/or quantity following nearby fracturing operations.

And fracking has also contributed to dangerous seismic activity. After an injection well near Youngstown, Ohio, was linked to earthquakes there, Governor John Kasich in 2012 issued an executive order requiring seismic studies before the state will issue well permits to operators. Yet despite the risks, no other state or the federal government requires any type of seismic risk assessment for injection wells.

Fracking alarmists have some statistics on their side. The average yearly rate of earthquakes above 3.0 on the Richter scale from 1967 to 2000 was a mere twenty-one, according to the U.S. Geological Survey (USGS). But as fracking grew more common, this number jumped to about one hundred per year between 2010 and 2012. Oklahoma recorded on average of fewer than six earthquakes a year between 1975 and 2008. More than a hundred 3.0 quakes were recorded in the state in only the first four months of 2014. Thirty quakes were reported in just the first two months of 2015.

On Tuesday, August 19, 2014, the Oklahoma Geology Survey (OGS) reported an unprecedented twenty earthquakes in Oklahoma on just that one day. "No documented cases of induced seismicity have ever come close to the current earthquake rates or the area over which the earthquakes are occurring," reported the OGS. The OGS defines "induced seismicity" as minor earthquakes caused by human activity that might disrupt existing fault lines, whether by fracking, mass removal mining, reservoir impoundment, or geothermal production.

A prime example of the dangers of fracking came when a magnitude 5.7 earthquake—the largest ever recorded in the state—shook Prague, Oklahoma, in November 2011. It was preceded by a 4.7 foreshock and followed by a 4.7 aftershock. The quake injured two persons, destroyed fourteen homes, and closed schools. Seismic activity from the quake was felt in seventeen states.

Seismologists were initially puzzled. They felt the only possible culprit was the Wilzetta Fault, a 320-million-year-old rift lurking between Prague and nearby Meeker. Officials with the United States Geological Survey (USGS) found its database indicated a zero possibility of ground motion from the fault. "This fault is like an extinct volcano. It should never have been active," commented Katie Keranen, an assistant professor of geophysics at the University of Oklahoma, who led a team placing some two dozen seismometers around the Prague area. After a yearlong study that included scientists from the USGS and Columbia University's Lamont-Doherty Earth Observatory, Keranen concluded, "Pretty much everybody who looks at our data accepts that these events were likely caused by injection." Oklahoma is not alone in experiencing a recent upsurge in seismic activity. Unexpected quakes also occurred in Arkansas, Colorado, Ohio, and Texas.

Many cities and states are now considering antifracking laws or zoning ordinances. However, in politics, money talks. Despite support from two-thirds of the state's voters, the California Senate in May 2014 let die a bill that would have placed a moratorium on fracking in the state until scientific studies evaluated all health and environmental effects for both onshore and offshore drilling. The bill was defeated when four Democrats joined all twelve Republicans senators in voting not to forward the bill. Three other Democrats abstained, preventing the moratorium from gaining a majority.

The oil industry, led by the Western States Petroleum Association (WSPA), spent nearly $1.5 million in just the first three months of 2014 to lobby against the bill. The group Californians Against Fracking estimated that oil lobbyists ultimately spent a total of $15 million to defeat the bill.

Big business and the oil lobby were elated at the bill's defeat, while Zack Malitz with the antifracking group Credo Action, said, "The overwhelming majority of Californians who support a moratorium on fracking will not stop fighting fracking and the public health risks, earthquakes, and climate change linked to this toxic extraction process." Like California, the states of Pennsylvania, Texas, and Colorado have also allowed widespread fracking without evaluating its potential impact. The rise of fracking directly endangers the livelihood of people living in many of those states, and is another effective tool employed by the globalists to reduce population.

CHANGING THE GAME

THE HISTORY OF PETROLEUM DEPENDENCE IS FILLED WITH UNETHIcal business practices and death. Safer, more fuel-efficient cars such as the Tucker, Ford's Edsel, and the DeLorean were largely suppressed by those in power. The old Los Angeles "Red Car" electric trolley system was put out of business by the freeway systems designed for petroleum-fueled vehicle traffic. The German inventor Rudolf Diesel originally designed his engine to run on organic matter, namely peanut oil. But he mysteriously died while crossing the English Channel in 1913, and today diesel engines run on various forms of distilled petroleum. Alternative fuels, derived from resources other than petroleum, include ethanol, produced domestically from corn and other crops; biodiesel, made from vegetable oils and animal fats; natural gas and propane, which produce less pollution; and hydrogen, which produces no pollution.

Other countries have made smart investments in sustainable transportation. In 2014, Japan celebrated fifty years of high-speed railroad, with trains traveling more than 200 mph. China built five thousand miles of high-speed rail in only six years. In the U.S., there is just one single high-speed rail line between Boston and Washington D.C., with another planned for the year 2025. Apparently, someone wants us to drive on those interstate highways, using all the gasoline we can afford.

The wholesale practice of fracking without adequate—and independent—studies of its impact on human health and water supplies is negligent and dangerous. Many states today are questioning the use of fracking, and many other nations have enacted moratoriums on fracking or even banned it outright. These countries include Canada, France, Germany, South Africa, Argentina, Ireland, Switzerland, and the Czech Republic. An informed citizenry must demand a more stringent review of fracking and its consequences to avoid future catastrophes from earthquakes and despoiled water.

Environmental degradation will continue until we invest in alternative energy. Thoughtful and concerted demands for safer and low-pollution technologies must increase until we can break the stranglehold the oil and gas industry holds over public policy.

CHAPTER 3

DRUGS

Despite draconian antidrug laws and harsh sentences for drug offenders, the drug problem has only worsened over the years.

According to the United Nations' "Drug Report 2012," the citizens of the United States take more drugs, both legal and illegal, than almost any other nation. And a recent World Health Organization (WHO) survey of legal and illegal drug use in seventeen countries showed Americans are the world's largest consumers of illegal drugs, particularly cocaine and marijuana. Americans were four times more likely to report using cocaine in their lifetime (16 percent) than the next-closest country, New Zealand (4 percent). The U.S. also reported the highest rate of marijuana use (42.4 percent).

The survey found that persons with higher incomes were also more likely to use both legal and illegal drugs. "The use of drugs seems to be a feature of more affluent countries," the survey noted. "The U.S., which has been driving much of the world's drug research and drug policy agenda, stands out with higher levels of use of alcohol, cocaine, and cannabis, despite punitive illegal drug policies, as well as (in many U.S. states) a higher minimum legal alcohol drinking age than many comparable developed countries."

Just as both politicians and the public had to rethink Prohibition in the early 1930s, many nations today are beginning to reconsider the so-called War on Drugs, a war now commonly acknowledged as lost.

According to a 2014 report by five Nobel Prize–winning economists, the battle to create a drug-free world is "misguided and counterproductive." Entitled "Ending the Drug Wars," the study was produced by the London School of Economics' IDEAS center, and posits that a fundamental restructuring of national and international policies and strategies is required.

"The [current] strategy has failed based on its own terms. Evidence shows that drug prices have been declining while purity has been increasing. This has been despite drastic increases in global enforcement spending. Continuing to spend vast resources on punitive enforcement-led policies, generally at the expense of proven public health policies, can no longer be justified," argued the eighty-two-page report.

According to this report, "The pursuit of a militarized and enforcement-led global 'war on drugs' strategy has produced enormous negative outcomes and collateral damage. These include mass incarceration in the U.S., highly repressive policies in Asia, vast corruption and political destabilization in Afghanistan and West Africa, immense violence in Latin America, an HIV epidemic in Russia, an acute global shortage of pain medication and the propagation of systematic human rights abuses around the world." It concludes, "It is time to end the 'War on Drugs' and massively redirect resources towards effective evidence-based policies underpinned by rigorous economic analysis."

Decrying a "one-size-fits-all" approach to combating illicit drugs, the authors of the report said the United Nations should take the lead in creating a "new cooperative international framework based on the fundamental acceptance that different policies will work for different countries and regions." Such a new drug policy should be "based on principles of public health, harm reduction, illicit market impact reduction, expanded access to essential medicines, minimization of problematic consumption, rigorously monitored regulatory experimentation and an unwavering commitment to principles of human rights."

The International Drug Policy Project coordinator for the IDEAS center, John Collins, commented, "The drug war's failure has been recognized by public health professionals, security experts, human rights authorities and now some of the world's most respected economists. Leaders need to recognize that toeing the line on current drug control strategies comes with extraordinary human and financial costs to their citizens and economies."

Interestingly enough, it is not illegal drugs such as heroin or cocaine that have proven most fatal to citizens of the death culture.

As much as a third of the population—one hundred million Americans—take powerful and even toxic mood-altering legal drugs just to get through the day. It is particularly scary to note that, according to some studies, some forty million of these drugged persons admit to driving on public thoroughfares while under the influence.

Pharmaceutical companies' profits have skyrocketed since the 1980s. From 1960 to 1980, prescription drug sales were fairly static as a percentage of U.S. gross domestic product, but between 1980 and 2000, they tripled. By 2012, prescription drug spending was estimated to account for $260.8 billion of national health spending.

The success of Big Pharma has more to do with marketing than with the efficiency of its drugs. Attempts by the large drug corporations to convince healthy people that they are sick and need drugs is called disease-mongering. To increase sales, Big Pharma invents new diseases, such as female sexual dysfunction syndrome, premenstrual dysphoric disorder, toenail fungus, male baldness, and social anxiety disorder (formerly known as shyness). These are but a few of the normal or mild conditions that have been represented as diseases requiring medication.

In a 2014 article in *Whistleblower* magazine, editor David Kupelian foresaw widespread and increasing drug use was bringing America a "real zombie apocalypse." Kupelian noted, "It turns out, whether we're caught up in the huge illegal drug world or the equally huge legal drug world—either way, tens of millions of us are taking basically the same drugs to deal with basically the same problems, and we're headed for basically the same dead end." He added, "Though psychiatry is supposed to be helping us, it has lost its way and become the most drug-dependent of all."

CHAPTER 4

DEADLY MEDICINE

IN 2000, THE *Journal of the American Medical Association* (*JAMA*) shocked the medical profession by revealing that the third leading cause of death for all Americans was doctor-related. A study published in the July 26, 2000, edition of *JAMA* by Dr. Barbara Starfield indicated that as many as 225,000 deaths per year are the result of conventional medical care. And some researchers say this is a conservative estimate; the real total may be closer to one million preventable deaths yearly . In either case, this makes the American medical system the third leading cause of death in the U.S., right behind heart disease and cancer.

Starfield, who died in 2011, was a much-honored pediatrician and cofounder of the International Society for Equity in Health. In her 2000 report, entitled "Is U.S. Health Really the Best in the World?," Starfield revealed that there were 12,000 yearly deaths from unnecessary surgeries, 7,000 deaths from medication errors in hospitals, 20,000 deaths from other errors in hospitals, and 80,000 deaths from infections acquired in hospitals. Another 106,000 deaths were attributed to FDA-approved and correctly prescribed medicines. This number does not include illegal drugs or the "inappropriate use" of legal prescription drugs.

Starfield, in a 2009 interview, declared, contrary to what most people think, that the U.S. does not have the best health in the world. "The American public appears to have been hoodwinked into believing that more interventions lead to better health," she said.

According to Starfield, most have never heard of deadly diseases such as carbapenem-resistant Enterobacteriaceae (CRE), which the World Health Organization (WHO) has labeled "one of the three greatest threats to human health." Today, this new, highly contagious, drug-resistant bacteria has a fatality rate as high as 50 percent and is being seen more and more often in long-term facilities rather than large hospitals, according to the CDC.

And we're helping to create these killer germs. CRE has become more prevalent as a result of doctors' overprescribing antibiotics using narrowly targeted chemical medications that lack the holistic benefits found in natural remedies. Such restrictive use encourages bacteria to develop molecular defenses, resulting in increased resistance to antibiotics. "The situation is so bad today that the entire pharmaceutical industry has no drug, no chemicals and no experimental medicines which can kill CRE superbugs," noted *Natural News* editor Mike Adams, adding, "Drug companies have discovered that it's far more profitable to sell 'lifestyle management' drugs like statin drugs and blood pressure drugs than to sell antibiotics which treat acute infections. Antibiotics simply aren't very profitable because relatively few people acquire such infections. Meanwhile, everyone can be convinced they might have high cholesterol and therefore need to take a statin drug for life." Were a superbug like CRE to gain widespread traction, America would be defenseless to stop its spread.

While natural health advocates recognize the value of allopathic drug treatment in certain cases, they argue that natural remedies along with proper diet and exercise can often aid the normal person in maintaining health.

Natural health advocate Dr. Joseph Mercola pointed out that Japan has benefited from the understanding that modern technology is wonderful, but just because it can be used to diagnose illnesses does not mean one should be committed to undergoing traditional treatment. Mercola noted that Japan's health statistics reflect that loving care is often more effective than treatment. Drugs, surgery, and hospitals are less effective than simple improvements in diet, exercise, and lifestyle.

There is no logical reason—apart from the profits enjoyed by the elites at the head of pharmaceutical companies—why our health care system should focus exclusively on these allopathic "cures." Some consider the greatest health conspiracy of all time to be the institutionalized effort to prevent the public from realizing that humans are born with all the programming they need to create perfect health and can often heal themselves of deadly diseases.

What's more, federal agencies have a woeful track record of protecting the public from highly dangerous drugs. Starfield explains: "Even though there will always be adverse events that cannot be anticipated, the fact is that more and more unsafe drugs are being approved for use. Many people attribute that to the fact that the pharmaceutical industry is (for the past ten years or so) required to pay the FDA [Food and Drug Administration] for reviews—which puts the FDA into an untenable position of working for the industry it is regulating." The death statistics Starfield reports are evidence of the broken U.S. health care industry. And the powerful forces behind it are unwilling to relinquish their grip on U.S. health policy. Insurance companies, specialty- and disease-oriented medical academia, and the pharmaceutical- and device-manufacturing industries all contribute heavily to congressional campaigns, often lobbying for policies not in the public's best interest.

This cozy relationship between the FDA, pharmaceutical companies, and Congress allows these deadly medicines to make their way onto the market. Even drugs thought to be safe can prove deadly, including over-the-counter remedies such as ibuprofen and acetaminophen, the active ingredient in Tylenol, one of the nation's most popular pain relievers. Even small overdosing of acetaminophen has been shown to cause long-term liver damage and in rare cases can cause stomach bleeding and even death. According to the CDC, about 150 Americans each year die from accidental acetaminophen overdoses.

The FDA and the makers of acetaminophen such as Johnson &

Johnson's Tylenol are well aware of risks associated with the drug, as indicated on the warning labels, but the FDA has yet to finalize rules governing its safe use. The FDA safety review of acetaminophen began in the 1970s, but by 2013, it was still incomplete because of foot-dragging within the agency. FDA records show the agency repeatedly defers consumer protection decisions even when they are endorsed by the agency's own advisory committees.

Although Johnson & Johnson has moved toward making Tylenol more safe, separately developing an antidote to acetaminophen poisoning, internal corporate documents obtained by the watchdog organization ProPublica show that for more than three decades the company has fought against safety warnings, dosage restrictions, and other measures designed to make the product safer.

FDA officials admit the agency has moved sluggishly to address the mounting toll of liver damage and deaths attributed to acetaminophen. Dr. Sandy Kweder with the FDA said, "Among over-the-counter medicines, it's among our top priorities. It just takes time." Informed consumers wonder if forty years is not enough.

Some deadly drugs are not only dangerous to your health but also tough on your pocketbook. Ever wonder why the costs of prescription drugs in the United States are the highest in the world or why it's illegal to import similar, cheaper drugs from Canada or Mexico?

This situation can be traced back to a Medicare prescription drug program pharmaceutical companies lobbied for the passage of in 2003. The Medicare Prescription Drug, Improvement, and Modernization Act, the largest overhaul of Medicare in its history to that date, extended limited prescription drug coverage under Medicare to forty-one million Americans, including thirteen million who had never before been covered, with an estimated cost of $400 billion.

Lobbyists for pharmaceutical companies, which spend roughly $100 million a year in campaign contributions and lobbying expenses, were present throughout the development of the bill, in fact, outnumbered Congress members two to one. The legislation, one of the most expensive bills ever placed before the House, was passed in an unorthodox roll-call vote in the middle of the night.

"The pharmaceutical lobbyists wrote the bill," recalled North

Carolina Republican Walter Jones. "The bill was over a thousand pages. And it got to the members of the House that morning, and we voted for it at about 3:00 a.m. I've been in politics for 22 years, and it was the ugliest night I have ever seen in 22 years." Critics claim the law provided billions of dollars in subsidies to insurance companies, giant pharmaceutical corporations, and health maintenance organizations, and took the first step toward allowing private plans to compete with Medicare.

Many of the most expensive drugs are used in the fight against cancer, the second leading cause of death in America. As the baby boomer generation reaches retirement age, medical expenditures for cancer research and treatment is expected to top $158 billion (in 2010 dollars) by the year 2020, according to the National Institutes of Health (NIH).

Chemotherapy, the use of chemical drugs to kill rapidly dividing cancerous cells, is responsible for many of the health care costs associated with cancer. Chemotherapy is a nasty procedure, which also kills normal cells, resulting in inflammation of the digestive tract, hair loss, and decreased production of blood cells, which suppresses the immune system.

One 2012 study found chemotherapy may actually cause more cancer than it cures. Chemo damages the DNA of healthy, noncancerous cells, triggering them to produce molecules that in turn produce more cancer cells. Researchers at the Fred Hutchison Cancer Research Center in Seattle found that chemotherapy damages healthy tissue surrounding tumors, causing cancer cells to develop resistance to the treatment. They transform into "super" cancer cells that resist chemotherapy, much as superbugs like CRE resist antibiotics, making the cancer even more deadly.

Typical of the growing hostility toward chemotherapy was this statement from Dave Mihalovic, a "naturopathic doctor" writing in *Waking Times:* "Ninety-seven percent of the time, chemotherapy does not work and continues to be used only for one reason—doctors and pharmaceutical companies make money from it. That's the only reason chemotherapy is still used." Mihalovic and other critics claim that chemotherapy, in actuality, boosts cancer growth and long-term mortality rates by destroying the immune system, increasing neurocognitive decline, disrupting endocrine functioning, and causing organ and metabolic toxicities. Patients basically live in a permanent state of disease until their death.

A twelve-year meta-analysis published in the *Journal of Clinical Oncology* shows that 97 percent of the time chemotherapy is ineffective in regressing metastatic cancers. One wonders where the money raised for cancer is actually going: certainly not to nutritional or homeopathic research, which are likely to be more effective than chemo has been. As Peter Glidden, a nutritional doctor and author of *The MD Emperor Has No Clothes,* remarks, "If Ford Motor Company made an automobile that exploded 97 percent of the time, would they still be in business? No." Yet the influence of pharmaceutical companies keeps the present broken system in place, despite its obvious failures.

If there were the money and political will to explore other treatment options, many seem to hold promise. Dr. Cristina Sanchez of Complutense University of Madrid has found that tetrahydrocannabinol (THC), the psychoactive chemical found in marijuana, kills cancer cells without damaging healthy cells. Her research, along with studies at the University of London's St. George's Medical School, show that THC has "potent anti-cancer activity," and can significantly "target and switch off" pathways that allow cancers to grow.

As of this writing, marijuana continues to be outlawed, even for research, as a Schedule I narcotic by the U.S. federal government, along with morphine, LSD, heroin, and opium. Could the cancer-killing properties of marijuana explain why the federal government continues to ban it?

One early cancer researcher who spelled out a cure was Nobel laureate physiologist Otto Warburg. Prior to World War II, Warburg gave a lecture describing both the cause and cure for cancer. "Summarized in a few words, the prime cause of cancer is the replacement of the respiration of oxygen in normal body cells by a fermentation of sugar. All normal body cells meet their energy needs by respiration of oxygen, whereas cancer cells meet their energy needs in great part by fermentation. All normal body cells are thus obligate aerobes, whereas all cancer cells are partial anaerobes . . . Oxygen gas, the donor of energy in plants and animals is dethroned in the cancer cells and replaced by an energy yielding reaction of the lowest living forms, namely, a fermentation of glucose," he announced, adding, "On the basis of anaerobiosis there is now a real chance to get rid of this terrible disease."

In other words, while most living cells require oxygen to live, cancer cells do well without oxygen, instead drawing energy from the fermentation of sugars. Cancer cells cannot survive in an oxygenated alkaline system. Furthermore, humans require a minimum of 22 percent oxygen in the air they breathe to maintain normal health. Most American cities are regularly below this minimum, and on so-called ozone alert days, the oxygen percentage often drops to 18 percent, well below the level necessary for good health.

The amount of sugar in the American diet is well documented, with obesity quickly becoming a national health crisis. If Dr. Warburg's appraisal is correct, it is astounding that nothing has been done to cure cancer in the intervening eight decades since his lecture. Despite this knowledge, the average American's diet today remains the most acidic (sugar-based) in the world. Perhaps this is because, as suspicious researchers have observed, more people are making a living off cancer than dying from it.

Though anticancer drugs are the most profitable subindustry for pharmaceutical companies, other drugs as well have generated huge profits while actively doing harm to those who take them. The antiinflammatory drug Vioxx is one such example. Before finally being withdrawn from the market in 2004, Vioxx was believed to have caused more than sixty thousand deaths. Merck, the producer of the drug, is the second largest pharmaceutical corporation in the U.S., and profited tremendously from Vioxx, which earned $2.5 billion in sales in 2003 alone. When the drug was pulled due in large part to evidence that it contributed to fatal heart attacks and strokes, analysts anticipated that the judgment against Merck could run up to $25 billion. Yet the plea bargain reached in 2012 resulted in a fine of only $321 million, a mere blip on Merck's bottom line.

Equally worrisome was the accusation that Merck's clinical study reports were developed by Merck but ultimately published under the names of reputable doctors and scientists. Dr. Joseph S. Ross, of New York's Mount Sinai School of Medicine, found these apparently ghostwritten research studies while reviewing case documents in the lawsuits against the company. Faced with irrefutable evidence, Merck admitted hiring outside professional writers to develop research-related documents, including for its $500 million Vioxx marketing campaign. With clinical studies on the drug being written by people on the company payroll, it's no wonder that the dangerous effects of the drug were not discovered until it was too late.

The Vioxx case highlights a broader issue with pharmaceutical advertising. With the introduction of direct-to-consumer drug advertising in the 1990s, the number of Americans on prescription drugs for life has ballooned, with the number of adults and children on one or more prescription medications rising by more than twelve million between 2001 and 2007. And the goal is no longer to get off these drugs, as with an antibiotic, but rather to continue taking them in perpetuity.

Health reporter Martha Rosenberg points out that Big Pharma hooks the U.S. public on prescriptions for life. Prescriptions once used only as needed for pain, anxiety, GERD (gastroesophageal reflux), asthma, mood problems, migraines and even erectile dysfunction, gout and retroviruses are now "full-time" medicines. "Instead of having body pain to be treated transiently, you are put on an antidepressant like Cymbalta or seizure drug like Lyrica or Neurontin indefinitely," she said.

Arianna Huffington, president and editor in chief of the Huffington Post Media Group, asked, "So why don't things ever change, even as the [legal drug] death toll mounts? As always, the answer can be found by following the money. The big pharmaceutical companies continue to be the 800-pound gorillas of American politics, their power stemming from a muscular combination of lobbying ($150 million a year), campaign contributions (close to $50 million doled out to federal candidates over the past four years), and powerful friends in very high places (Donald Rumsfeld was formerly CEO of drug industry powerhouse G. D. Searle; and Mitch Daniels, the former White

House budget director and new governor-elect of Indiana, was a senior vice president at Eli Lilly.) . . . Of course, the real shame is that we continue to have a regulatory system in which corporate greed, political timidity and a culture of cronyism have rendered the public good a quaint afterthought."

Even scarier than the drugs we have now are the ones that may be to come. Scientists in the U.S. and other countries are creating new and previously unknown viruses that could wipe out whole populations. A report issued in May 2014 by epidemiologists Marc Lipsitch of the Harvard School of Public Health and Alison Galvani of Yale, noted labs around the world are creating and altering viruses to understand how natural strains might evolve into more lethal forms. These researchers have warned that the benefits of such experiments might be outweighed by the risk of pathogenic strains escaping from laboratories and spreading.

The U.S. government, which funds many of these controversial experiments, instituted new rules that fund such work provided the potential benefits are deemed substantial and the risks considered manageable. Lipsitch argues there is no evidence that the risks and benefits have been weighed properly. "To my knowledge, no such thing has been done, but funding for these experiments continues," he said. Many people fear such experimentation might lead to a pandemic, or fall into the hands of bioterrorists. Both the government and funding organizations should employ unbiased experts to assess which viral studies to support, yet this is not happening. With the recent spread of infectious diseases such as Ebola, this issue has been thrust into the public consciousness, and it's clear the U.S. government lacks a comprehensive system to determine which activities are safe and which are not.

CHAPTER 5

DEADLY VACCINES

IN THE EARLY 1950S, U.S. CHILDREN TYPICALLY RECEIVED THIRTEEN doses of four vaccines—for diphtheria, tetanus, pertussis, and smallpox—but not more than three doses in a single visit. By the mid-1980s, four more vaccines were added: measles, mumps, rubella (MMR), and polio. Today, the number of vaccines recommended by the Centers for Disease Control (CDC) is higher than in any country in the world. The CDC recommends children receive as many as thirty-seven doses of fourteen vaccines by the age of two and forty-eight doses of fourteen vaccines by age six, with sometimes as many as eight vaccine shots in a single visit. Critics point out that overvaccination, and the infusion into vaccines of additives such as fluoride, have had disastrous results for children. The battle lines of the vaccine debate are clear: independent doctors and researchers have lined up against government regulators and the pharmaceutical industry.

Another critic is Dr. Lawrence Palevsky, a board-certified pediatrician trained at the New York School of Medicine. He explained his transition from vaccine believer to vaccine skeptic: "When I went through medical school, I was taught that vaccines were completely safe and completely effective, and I had no reason to believe otherwise. All the information that I was taught was pretty standard in all the medical schools and the teachings and scientific literature throughout the country. I had no reason to disbelieve it . . . But more and more, I kept seeing that my experience of the world, my experience in using and reading about vaccines, and hearing what parents were saying about vaccines were very different from what I was taught in medical school and my residency training."

Palevsky, along with other physicians, has become concerned that vaccines have not been completely proven safe or even completely effective, based on the literature now available.

Yet many doctors remain unwilling to consider the possibility that something might be wrong with vaccines. "Most pediatricians are indoctrinated to simply tell parents that anything related to a bad outcome from a vaccine is a mere coincidence. But how come there are so many of these coincidences?" Palevsky wonders. "It is heartbreaking, because I see many of these kids who were developmentally normal, who were doing well, who were speaking, then whose voices and eye contacts were lost, who went into seizures, who developed asthma and allergies, and they had nowhere to go because their doctors told them that they don't know what they're talking about. These kids are real."

Palevsky argues that proper vaccine studies have not been conducted and that medical literature is "pretty supportive of the fact" that vaccines have "much greater adverse outcomes on the genotype of the body, the immune system of the body, the brain of the body, and the intracellular functions of the body than we are willing to tell the public about."

Despite proponents' claims that more than twenty thousand studies have proven the safety of vaccines, a closer look reveals otherwise. Before such studies can be uncritically accepted, it should be asked how the study was designed and by whom. Were there any control groups? What was the size of the study's population and were they kids or adults? "I think we will see that in most of those studies, the actual safety has never really been proven," opined Palevsky.

Proponents of vaccines also argue that unvaccinated people benefit from the fact that most children are vaccinated. This is called herd immunity: the idea is that the more people that are vaccinated, the less likely it is that anyone in that community will become infected.

This has been cast into serious doubt by members of the medical community. Dr. Palevsky and others believe it is preposterous to think that children who are vaccinated no longer carry the bacteria or the viruses for which they have been inoculated. If, in fact, children are vaccinated, then why should parents and public health authorities be afraid that nonvaccinated children are somehow carrying something that their children are not. "You can't have it both ways," argued Dr. Palevsky. "You can't vaccinate believing that your children are protected and then feel that your children are not protected because somehow, some non-vaccinated child is carrying some secret organism that no one else is carrying. It just doesn't make any sense."

Others agree. Dr. Russell Blaylock writes, "That vaccine-induced herd immunity is mostly myth can be proven quite simply. When I was in medical school, we were taught that all of the childhood vaccines lasted a lifetime. This thinking existed for over 70 years. It was not until relatively recently that it was discovered that most of these vaccines lost their effectiveness 2 to 10 years after being given. What this means is that at least half the population, that is the baby boomers, have had no vaccine-induced immunity against any of these diseases for which they had been vaccinated very early in life."

In essence, this would mean at least 50 percent or more of the population lost their immune protection within two to ten years of being vaccinated, thus, most of the population today goes without the 95 percent number claimed needed for herd immunity.

Most medical authorities agree that the vaccine question does not lend itself to simple yes or no answers. Should parents forbid any vaccination whatsoever or allow whatever their doctor prescribes? More and more Americans are deciding that such questions should be left up to the individual parent. Proponents of informed consent contend that families who have done their homework should be able to make their own choices. "When parents are given both sides, it is up to them to make that informed choice," advised Dr. Palevsky.

What is clear, however, from the history of vaccination is that there are very real risks associated with the practice. In 1998, three vaccine manufacturers faced lawsuits in the UK after parents there reported more than 1,500 instances of side effects following the administration of measles, mumps, and rubella (MMR) vaccines. Despite assurances from British health officials that there was no connection between the reported symptoms and the vaccines, cases were set for trial in the High Court to decide whether the vaccines caused symptoms of autism and bowel problems. The cases were funded under the English legal aid system and supported by twenty-seven experts who prepared reports supporting the children's cases. The parents believed their children were normal before being vaccinated, and saw nothing but the vaccinations to account for the subsequent changes. The cases stalled and have not proceeded after legal aid was withdrawn in August 2003, leaving the question of vaccine culpability unresolved.

While supporters claim vaccinations are necessary for good health, an article published in *Pediatrics Journal* refuted this claim. It described how the majority of infectious diseases were reduced prior to World War II thanks to greater health care and nutrition, better sanitation and improved living conditions. This was long before the widespread use of antibiotics and vaccinations.

"During the past 30 years, the number of vaccinations our children receive has tripled, and during that same period of time, the number of children with learning disabilities, ADHD, asthma and diabetes has also more than tripled!" noted Dr. Joseph Mercola. "It's astounding how effective drug companies are at manipulating national health policy. They have been able to manipulate and pervert the public health system so they now can sell toxic, ineffective flu shots in pharmacies, airports, college campuses, grocery stores, and countless other outlets, without *any* [emphasis in the original] solid evidence that the flu shots even work, and despite the many questionable ingredients found in the shots."

Such ingredients include thimerosal, known to be very toxic by inhalation, ingestion, and contact with skin yet still included in multidose flu vaccine despite a legitimate public outcry. Other questionable vaccine ingredients include formaldehyde or formalin, chicken kidney cells and DNA, monosodium glutamate, Octoxynol-9 (Triton X-100), used in spermicides, and polysorbate 80, an emulsifier found to cause an anaphylactic reaction in a German patient.

Furthermore, many vaccines are making recipients vulnerable to new and dangerous diseases. In 2014, a severe respiratory virus known as enterovirus D68 (EV-D68) infected more than 150 children in the midwest. By the fall, victims, many of whom required intensive care, were reported as far north as Canada. Dr. Mary Anne Jackson, division director of infectious diseases at Children's Mercy Hospital in Kansas City, Missouri, said, "It's worse in terms of scope of critically ill children who require intensive care. I would call it unprecedented. I've practiced for 30 years in pediatrics, and I've never seen anything quite like this." The only common denominator of those infected with D68 was that all had been vaccinated for MMR, influenza, and polio, indicating this could be an unintended side effect.

Yet despite the inherent risk posed by vaccines, they are often used as a political tool to capitalize on public fear. The threat posed by a number of pandemics in recent years is evidence of this. Although CDC predictions as to the scope and danger of a pandemic are frequently inaccurate, the government still uses these predictions to push an agenda of increased vaccinations onto the citizenry. For example, the World Health Organization (WHO) and the CDC in 2009 predicted a pandemic of swine flu, a combination of human influenza viruses, avian (bird) flu, and swine flu strains. The WHO and the CDC, in predicting a serious swine flu epidemic warned that compulsory inoculations might be necessary.

At the time, there were indeed reports of employees losing their jobs if they refused to be vaccinated. One nurse fought back and won . . . after four years. June Valent, a nurse at Hackettstown Community Hospital in New Jersey, was fired for refusing the flu shot. The hospital required employees to take the flu vaccine unless they had a medical or religious excuse, and anyone who refused was required to wear a mask. Valent agreed to wear a mask but declined the shot. As a result, she was fired and disqualified for unemployment benefits. She filed a lawsuit and in 2014 a court agreed that that the hospital violated Valent's right to freedom of expression by accepting a religious-based exemption while denying her secular one.

Given the ultimately limited reach of the outbreak, and the number of people who were vaccinated, it's fair to wonder whether pharmaceutical corporations may have manipulated the WHO in an effort to sell swine flu vaccine. These companies had invested millions of dollars researching and developing pandemic vaccines following the bird flu scares of 2006 and 2007; without a pandemic, either real or manufactured, this money would have gone to waste. Whether these corporations and the WHO deliberately misled the public or not, they clearly overstated the potential dangers of the flu strain. As Dr. Wolfgang Wodarg, the president of the Health Committee of the Council of Europe, puts it, the pharmaceutical lobby and world governments engaged in a "great campaign of panic" centered on the swine flu.

Such suspicions gain currency when paired with incidents like the following: in September 2014, the European Centre for Disease Control and Prevention (ECDC) revealed that the pharmaceutical giant GlaxoSmithKline had released more than eleven gallons of concentrated live polio virus into rivers near Rixen, Belgium. The release, termed accidental, if in fact purposeful, would demonstrate a reliable way to ensure that the polio vaccine continues to be in high demand. Pharmaceutical companies also benefit from the interactions between multiple different vaccines. A 2009 Canadian study indicated people under the age of fifty are at a higher risk of being infected with the swine flu virus after receiving the annual flu shot. Lead authors Danuta Skowronski of the British Columbia Centre for Disease Control and Gaston De Serres of Laval University, found consistency across four epidemiologic studies and one animal experiment that suggested "an association that cannot be dismissed on the basis of chance and is unlikely to be explained entirely by bias." Canadian authorities expressed concern the results of this study might throw inoculation programs into disarray with fewer people volunteering for shots.

And when one fake pandemic fizzles out, the government-pharmaceutical complex simply moves on to the next. When fears over swine flu waned, the media began touting a new pandemic known as Middle East Respiratory Syndrome (MERS), which was first reported in Saudi Arabia in 2012 and then spread westward. By mid-2014, at least two cases of MERS were reported in the U.S., both in persons who had traveled there from Saudi Arabia. Although the CDC warned that MERS carries a 30 percent death rate for sufferers, officials said the viral respiratory disease is contracted only through close contact and was not likely to spread widely.

When they're not playing up fears of certain diseases in order to inject us with deadly vaccines, the pharmaceutical establishment is downplaying the risks posed by legitimately deadly diseases. In 2014, of course, the hottest disease-promoting fear of depopulation was the Ebola virus, spreading rapidly in several West African countries. The arrival in the U.S. of the first American to contract Ebola, Dr. Kent Brantly, prompted fears of a pandemic, with many questioning the wisdom of intentionally bringing Ebola victims to the U.S. The Ebola virus was spreading faster than it could be controlled and had the potential to be "catastrophic" in terms of lost lives, warned Margaret Chan, head of the World Health Organization (WHO). So why then were victims being brought to the United States?

The immediate response by African governments was equally incoherent. In the early fall of 2014, the West African nation of Sierra Leone, situated between Liberia and Guinea, was placed under medical martial law, with its six million citizens quarantined in their homes for three days as government workers went door to door seeking Ebola victims.

Similarly, in Liberia, the government and the World Health Organization were establishing about seventy "community care centers" to involuntarily move those infected with Ebola from their homes. A similar program was planned for Sierra Leone. Some observers have called these controversial centers "Ebola death camps." At least three thousand U.S. soldiers were ordered to aid the lockdown and displacement, causing some Americans to suspect the African action might be a preview of things to come in the U.S. should the Ebola outbreak spread here. Oddly enough, at least some of these troops were assigned hospice duties requiring close contact with the infected, such as cleaning rooms and emptying bedpans.

By the end of September 2014, the first Ebola case in the USA was made public by the CDC. The critically ill patient, who was not initially named, entered Presbyterian Hospital in Dallas after arriving in the U.S. from Liberia to visit relatives. CDC Director Thomas Frieden appeared optimistic when he announced, "This is the first case of Ebola diagnosed in the U.S. and the first strain of this Ebola diagnosed outside of Africa. I have no doubt that we will control this case of Ebola so that it does not spread widely in this country."

Yet it appeared the federal government was gearing up for an Ebola pandemic following the activities in Africa when in September 2014 President Obama warned that the Ebola outbreak was "spiraling out of control" and that the world might see "hundreds of thousands of people infected" if it was not swiftly addressed by the international community. Meanwhile, Lakeland Industries, a manufacturer of protective clothing, announced the U.S. State Department had put out a bid for 160,000 hazmat suits for the battle against Ebola.

Thus far, the early indications of what the Obama administration response to Ebola may be are troubling. Soon after the outbreak, Obama signed an executive order authorizing the apprehension, detention, or conditional release of individuals with certain severe acute respiratory diseases, not including influenza. This order modified a similar one signed by George Bush in 2003 in response to public fear over the SARS epidemic.

This order could be used to forcibly quarantine persons merely suspected of carrying the Ebola virus. The CDC acknowledged that twenty quarantine centers, scattered across the country, had been activated to accept patients. The contradiction between the government's avowed actions to protect the country from Ebola and the Obama administration's lenient immigration policy, which has allowed a flood of illegals carrying any number of diseases across the nation's southern border, has been noted by many.

Kurt Nimmo, writing for Prison Planet.com, also noted this contradiction and claimed, "Disease, natural disaster, and man-made crises are routinely exploited by government as pretexts to enlarge and extend its power and reach. The state and its propaganda media thrive on one manufactured crisis after another as part of a systematic effort to ramp up the police state."

The apparent goal is not protection of the people but the creation of an all-encompassing surveillance state along with a militarized component designed to control the population through fear and violence. By April 2015, at least half of the American troops sent to Africa had returned and the Ebola outbreak appeared to have waned, but only after almost 10,000 deaths.

Ebola is difficult to contract, as it is primarily passed only by direct contact with blood and other bodily fluids from an infected person. Yet seizing on irrational public fear, the government has been successful in imposing new restrictions on citizens. In addition to providing an excuse for an enlargement of government and military power, hyped-up pandemic threats are used by transnational corporations to increase profits. The fearmongering over Ebola is simply the same swine flu drama played out on a grander stage.

Then there is the question of why the U.S. government has claimed ownership of all Ebola viruses that share more than 70 percent similarity with an Ebola strain that was patented in 2010 (patent No. CA2741523A1). One of the listed patent's applicants was "The Government of The United States of America as represented by the Secretary, Department Of Health & Human Services, Center For Disease Control."

With all of these created pandemics, we must ask what the ultimate end goal might be. In this case the answer is clear: they are key components of the globalist agenda of population reduction. The scientific community has long suggested the Ebola virus could be an effective tool toward this end. In 2006, Eric Pianka, a professor of biology at the University of Texas, was reported to have advocated severe population reduction by means of a virus such as Ebola. Pianka has since disputed the reports and clarified that he does "not bear any ill will toward people," but at least one first-hand account has confirmed the story. Forrest Mims, writing in the *Citizen Scientist*, noted, "I watched in amazement as a few hundred members of the Texas Academy of Science rose to their feet and gave a standing ovation to a speech that enthusiastically advocated the elimination of 90 percent of Earth's population by airborne Ebola."

The text of Pianka's speech is chilling. He spoke of a future with a drastically reduced population, noting that, "After the human population collapses, there's going to be a lot fewer of us. Food's going to be diminished. Pollution's going to go down, which will be good," Pianka said. He also suggested that the entire population of the earth be sterilized. And a pandemic virus such as Ebola, capable of wiping out a huge percentage of the human population, is a key element of Pianka's disturbing vision.

As if the disease is not deadly enough on its own, the military has also been brought into play in the Ebola epidemic. The Pentagon, in addition to the troops sent to Africa, has deployed biological diagnostic systems to National Guard support teams in all fifty states in the event of any national emergency event, which could include a pandemic outbreak, even though they claim such an event is improbable. Why then are so many troops stationed across the country?

Can it be that people who rely solely on Western allopathic medicine will die en masse, victims of a medical system that suppresses medical alternatives offering effective remedies from pandemics? It is quite remarkable that even as Western medical professionals admit to having no guaranteed remedy against a deadly supergerm, they nevertheless dismiss any homeopathic treatment regardless of its efficiency and don't want people to have access to anything else that might work either. They echo corporate pharmaceutical officials who warn that herbal medicine might be dangerous . . . never mind the deadly supergerm.

No one argues that Western medicine should not be used in a pandemic, only that every possibility, including both Western and Eastern medicine, herbal medicine, nutritional therapies, sunlight therapy, should be considered. Only a fool would limit his options to the one single system of medicine that admittedly offers no cure.

With no cure-all forthcoming from medical science, it is perhaps time to question the effectiveness and hazards of vaccinations.

In many cases, parents have reported abnormal reactions following vaccinations. These included seizures, spiking fevers, diarrhea, lethargy, high-pitched screaming, and other abnormalities. The damage may be coming not from the vaccines themselves but from added ingredients such as thimerosal, a preservative used by vaccine manufacturers to save money on production and storage. Thermerosal, which contains 49.6 percent mercury by weight, can metabolize or degrade into ethylmercury.

A paper published in the June 2009 issue of *Toxicological and Environmental Chemistry* (volume 91, issue 4) concluded the routine administering of childhood vaccines containing thimerosal could cause "significant cellular toxicity in human neuronal and fetal cells."

"This latest study confirms that damage [from thimerosal] does occur in human neuronal and fetal cells, even at low concentrations," wrote Dr. Joseph Mercola of the Illinois Natural Health Center. He noted that rates of autism in the U.S. have increased nearly sixty-fold since the late 1970s, right along with the increasing number of vaccinations added to the childhood vaccination schedule. Typically, by the age of three, most children have received at least twenty-four of their scheduled vaccinations. It is at this stage that symptoms of autism may become apparent.

The academic community's perspective on thimerosal is indeed grim. As Boyd Haley, a retired professor of chemistry at the University of Kentucky, noted, "If you inject thimerosal into an animal, its brain will sicken. If you apply it to living tissue, the cells die. If you put it in a petri dish the culture dies. Knowing these things, it would be shocking if one could inject it into an infant without causing damage."

And we should be doubly suspicious of these additives given the remarkably poor track record of pharmaceutical companies when it comes to ethics and safety. In 2012, GlaxoSmithKlein, while denying any wrongdoing, nevertheless pleaded guilty to criminal charges and paid a $3 billion fine for promoting its antidepressants for unapproved uses and failing to report safety data about a top diabetes drug. It was the largest settlement recorded involving a pharmaceutical company. The agreement also included civil penalties for improper marketing of a half-dozen other drugs.

Prosecutors claimed Glaxo used several tactics aimed at promoting the drug Paxil in children when the FDA had not approved it for non-adults. These included arranging the publication of a medical journal article that misreported data from a clinical trial.

With such a sketchy history, how can we trust pharmaceutical companies when they exhort us to inject newborn babies with potentially harmful vaccinations?

In part, the problem is that no individual employees are ever held accountable for these sorts of deadly oversights. No individuals were charged in the GlaxoSmithKline case, nor in the many criminal suits involving other drug manufacturers. High-dollar settlements were the order of the day, as recorded against Abbott Laboratories, Johnson & Johnson, and others.

But large fines don't appear to affect the questionable practices of pharmaceutical companies; such practices will not change until drug executives are singled out for punishment. Former New York attorney general Eliot Spitzer, who sued GlaxoSmithKline in 2004, remarked, "What we're learning is that money doesn't deter corporate malfeasance. The only thing that will work in my view is CEOs and officials being forced to resign and individual culpability being enforced."

We could learn from the promising prosecution of Dong-Pyou Han, former laboratory manager at Iowa State University. In June 2014, federal prosecutors actually charged Han, a vaccine researcher, with crimes involving making false statements. Han confessed to adulterating rabbit blood with human antibodies to create the impression that an experimental HIV vaccine might be working. After $5 million was spent in National Institutes of Health (NIH) grants, it was found that Han's results, once hailed as "groundbreaking," were fraudulent. In February 2015, Han accepted a plea deal under which he faces prison time and restitution payments. Han's case illustrates the shortcuts and downright fabrications of some vaccine researchers. Vaccines have been especially in the public eye recently because of growing concerns over the possibility that multiple vaccinations may lie behind the rising incidence of autism.

Autism is a real concern for parents wondering about vaccines.

In March 2014, the CDC reported that one in every sixty-eight U.S. children has an autism spectrum disorder (ASD). This is a 30 percent increase from one in eighty-eight only two years previously. This newest estimate is based on the CDC's evaluation of the health and educational records of all eight-year-old children in eleven states.

Even more disturbing is the claim by a former CDC scientist that the center, under the leadership of Dr. Julie Gerberding, committed fraud by altering study data that established a link between vaccination and autism. The study in question indicated that mumps, measles, and rubella (MMR) vaccinations caused a 350 percent increase in autism in black children. Dr. William Thompson, who conducted major research on the MMR vaccine, said CDC officials knew of this danger as far back as 2001.

In support of his theory that the CDC hid information on the vaccine was this public statement by Thompson in late August 2014: "I regret that my coauthors and I omitted statistically significant information in our 2004 article published in the journal *Pediatrics.* The omitted data suggested that African American males who received the MMR vaccine before age 36 months were at increased risk for autism. Decisions were made regarding which findings to report after the data were collected, and I believe that the final study protocol was not followed."

Dr. Thompson further stated, "The CDC has put the [autism] research ten years behind. Because the CDC has not been transparent, we've missed ten years of research [on the autism-vaccine connection] . . . CDC is . . . they're paralyzed. The whole system is paralyzed right now . . . I have a boss who's asking me to lie . . . if I'm forced to testify, I'm not gonna lie. I basically have stopped lying."

When vaccine researchers aren't outright lying over vaccine study data, pharmaceutical companies are accused of subtly manipulating the studies. In the fall of 2014, two former virologists for Merck, Stephen Krahling and Joan Wlochowski, added to the vaccine controversy when they filed a lawsuit against their former employer, claiming Merck defrauded the federal government by selling a vaccine that was less effective than advertised. The pair accused Merck of fraud by testing the effectiveness of its MMR vaccine against the version of the virus in the vaccine rather than against the natural mumps virus a person would catch in the real world, and of deceptively adding animal antibodies to the test, giving the false appearance of strong human immune response to the vaccine.

Attorneys for Merck asked for a dismissal of the case primarily because they claimed the FDA was the only authority entitled to act on vaccine issues. On September 5, 2014, Judge C. Darnell Jones II of the Federal District Court for the Eastern District of Pennsylvania mostly rejected Merck's motion to dismiss. Both opponents and proponents of vaccines are awaiting the results of this lawsuit.

Despite what pharmaceutical companies would lead us to believe, the evidence of a link between vaccines and autism is mounting despite the CDC's position that "numerous studies have found no association between thimerosal exposure and autism." Many children with autism had followed a normal development path before regressing just as they were receiving multiple vaccines during regular doctor visits. Health officials say the timing is purely coincidental. However, mercury and other heavy-metal poisons such as thimerasol appear to be a primary driver of autism, according to a 2014 study published in the journal *PLOS Computational Biology* by researchers from the University of Chicago. While they did not study a causal link between vaccines and autism, they found environmental pollution, and particularly mercury and mercury-containing compounds like those from coal-fired power plants and diesel engines, may increase a child's risk of developing an autism spectrum disorder (ASD) by nearly threefold.

Infants and young children are particularly vulnerable to mercury, a potent neurotoxin that can damage the brain, liver, kidneys, and central nervous system. Even small amounts of mercury are capable of causing a number of health problems, including impaired motor functioning, decreased cognitive ability, and emotional instability. Higher or prolonged exposure can result in even more serious health problems.

Other dangerous chemicals are also potentially responsible for causing autism. In addition to mercury, plasticizer chemicals, prescription drugs, and environmental pesticides contain tiny molecules that interfere with normal childhood development, leading to autism symptoms, according to the Chicago study. As the lead author of the study explains, "Essentially what happens is during pregnancy there are certain sensitive periods where the fetus is very vulnerable to a range of small molecules. Some of these small molecules essentially alter normal development. It's not really well known why, but it's an experimental observation."

The air we breathe may also be contributing to the autism epidemic. Exposure to traffic fumes, industrial air pollution, and other environmental toxins have all been shown to dramatically increase a mother's chances of having a child with autism. Researchers studied the insurance claims of about a hundred million U.S. citizens and used reports of congenital malformation in newborn boys as an indicator of parental exposure to environmental toxins. Pregnant women living in the top 20 percent of the most polluted areas evaluated were found to be twice as likely as women in the least polluted areas to birth a child with autism.

Autism isn't the only disturbing result of these dangerous chemicals. Evidence continues to mount that environmental factors, particularly heavy metals, may also play a role in the rise of many other diseases and neurological problems. A 1977 Russian study found that adults exposed to ethylmercury, the form of mercury in thimerosal, were at a much higher risk of brain damage later in life. Mercury intake has also been linked to cardiovascular disease, seizures, mental retardation, hyperactivity, dyslexia, and other debilitating conditions. As a result of these findings, Russia banned thimerosal from children's vaccines in 1980. Denmark, Austria, Japan, Great Britain, and all the Scandinavian countries have also banned the preservative. Yet thimerosal continues to be used as a key ingredient in some vaccines in the United States, and mercury can also still be found in dental amalgam fillings and some commercial fish products.

Aluminum is another dangerous element of vaccines. Aluminum is added to vaccines as an adjuvant in the belief that it will produce a stronger antibody response and thus be more protective. Aluminum is used in vaccines to selectively target T helper cells (Th1 and Th2), which push the immunization system to produce antibodies.

According to author and lecturer Dr. Lawrence Palevsky, who advocates a holistic approach to children's health, aluminum forces the undeveloped and immature immune system of infants and children to produce greater amounts of humoral immune cells (TH2) and antibodies, before their immune systems have a chance to adapt to the world in which they've just begun to live. "Under these circumstances, the activity of aluminum appears to play a vital role in disrupting the maturation of the immune system in infants and children through its effects on TH2," he wrote. Looking at the role of aluminum in vaccines, either acting alone or in conjunction with other vaccine ingredients and environmental toxins, Palvesky wonders what role this might play in the development of chronic illnesses in susceptible children.

Polysorbate 80, an emulsifying agent, is another potentially deadly additive in vaccines. Clinical studies have shown that polysorbate 80 increases the risk of serious side effects, such as blood clots, stroke, heart attack, heart failure, and, in some cases, death.

"If we're going to do justice to the topic of vaccine ingredients, we need to look at the potential harm of *all* [emphasis in the original] the vaccine ingredients at once, and examine their individual effects on our children's immune and nervous systems. Then, we can examine the interactive effects of the vaccine ingredients on human tissue, and evaluate the potential for harm . . . How many more children need to be potentially harmed before we invoke the precautionary principle of the Hippocratic Oath—First, Do No Harm?" asked Dr. Palevsky.

With the evidence growing that a number of the ingredients in vaccines may be deadly, even some politicians have expressed concern. Former representative Dan Burton of Indiana once asked the FDA to recall all vaccines containing the preservative thimerosal, explaining, "My only grandson became autistic right before my eyes—shortly after receiving his federally recommended and state-mandated vaccines. Without a full explanation of what was in the shots being given, my talkative, playful, outgoing healthy grandson Christian was subjected to very high levels of mercury through his vaccines. He also received the MMR vaccine. Within a few days he was showing signs of autism." Citing Canadian research concerning the toxicity of mercury, Burton noted, "We have gone from one in 10,000 children with autism to one in 88. It is worse than an epidemic, it is an absolute disaster."

Burton, while chairman of the Committee on Government Reform, asked that $128 million be appropriated to study the link between vaccines and autism. "Giving more money to research is not the only answer though," said Burton. "Oversight is needed to make sure that research that is funded will sufficiently answer the questions regarding the epidemic, how to treat autism, and how to prevent the next ten years from seeing the statistic of 1 in 250 from becoming 1 in 25 children. High quality clinical and laboratory research is needed now, not five or ten years from now." While the CDC has denied any connection between vaccines and autism, Burton urged independent analysis of old CDC studies. But this has yet to occur.

The stories of four prominent doctors connected to the autism debate—Brian Hooker, Andrew Wakefield, Thomas Verstraeten, and Poul Thorsen—help to explain the powerful forces that suppress criticism of vaccinations.

Brian Hooker holds a doctorate degree in biochemical engineering and has a teenage son with autism. Motivated by his son's condition and with the help of two congressmen, Hooker spent almost ten years submitting over one hundred Freedom of Information Act [FOIA] requests to the CDC for data linking mercury-filled thimerosal in vaccines to various disorders. He received thousands of documents, but with many key components blacked out. These documents included five CDC studies on thimerosal and autism written prior to 2004 that rejected the connection between thimerosal and autism.

Yet the heavily redacted documents Hooker received revealed the CDC had access to data linking thimerosal in vaccines to autism, nonorganic sleep disorders, and speech disorders but had concealed this from the public. Today, flu shots containing thimerosal still are administered to pregnant women and infants.

According to Hooker, a 2009 study hid data regarding the only valid part of the study (i.e., prenatal thimerosal exposure), which showed that children exposed to just sixteen micrograms of mercury in thimerosal in utero were up to eight times more likely to receive a diagnosis of regressive autism.

Hooker said his FOIA requests specifically sought information on five CDC studies on thimerosal and autism prior to 2004. These studies led to an Institute of Medicine (IOM) Immunization Safety Review Committee report, "Vaccines and Autism," released in May 2004. Denying any link between thimerosal in the MMR vaccine and autism, this report effectively shut down government funding for any independent research. "Given the constant reference that the CDC and others make to the 2004 IOM report, most of the key components of the FOIAed information have been completely redacted by the CDC," said Hooker, adding that much of the information sought in his FOIA requests has yet to be released by the CDC.

"I would challenge anyone who would rely on the veracity of the CDC studies," said Hooker. "They've repeatedly, purposefully withheld data that clearly show a link between thimerosal and autism (among other NDDs [neurodevelopmental disorders]). They've obfuscated the main issue via obviously biased statistical manipulation. Clearly, the CDC's conflicted role of vaccine advocate and vaccine safety guardian has contributed to this whole problem."

Hooker also noted that other countries that vaccinate less and have banned thimerosal have not experienced comparable autism rates to the United States. He added his belief that thimerosal has not been removed from U.S. vaccines because of "various issues in a concerted effort toward the globalization of vaccines."

Andrew Wakefield, a British former surgeon and medical researcher, supported Hooker in his effort to find the truth. Wakefield is a controversial figure who has been castigated by the medical establishment and the mainstream media. His Wikipedia page states that the doctor is "known for his fraudulent 1998 research paper in support of the now-discredited claim that there is a link between the administration of the measles, mumps, and rubella (MMR) vaccine and autism and other ailments." And in 2010, a five-member tribunal of the British General Medical Council (GMC) found Wakefield guilty of dishonesty and twelve counts involving the abuse of developmentally challenged children. He was barred from practicing medicine and the British medical journal *The Lancet* retracted his 1998 paper.

Yet Wakefield's claims are not nearly as outlandish as they have been made to seem by the media. In late 2011, the Strategic Autism Initiative (SAI), an autism research foundation, announced an investigation into Wakefield's claims headed by Dr. David Lewis, former senior-level research microbiologist for the U.S. Environmental Protection Agency (EPA) and a member of the National Whistleblowers Center (NWC) board of directors. The SAI rejected the assertion that Wakefield's claim was fraudulent. Further research has continued to support the doctor. More than twenty-six studies in the U.S., Canada, Venezuela, and Italy replicated Wakefield's 1998 findings but have not been widely publicized.

As director of the MRC's Research Misconduct Project, Lewis explained that important areas of scientific research are increasingly being manipulated by government agencies, large corporations, and leading universities in order to promote and protect their own interests. Suppressing independent research that threatens their interests is critical to their interests. While most scientists are rarely targeted for retribution by government, industry, or academia, some have faced false allegations of research misconduct. "Scientists who are targeted in this manner suffer lifelong consequences, and the chilling effect it has on other scientists is profound. Few, if any, scientists are willing to step into the firing line of government or big industry and risk being martyred," explained Lewis.

The outside pressure the pharmaceutical industry brings to bear on vaccine advocates explains why many doctors are wary of making any controversial statements about the effects of vaccines. Dr. Thomas Verstraeten is one such case. Verstraeten entered the vaccine fray when he authored a 2001 study whose initial phase seemed to indicate a potential link between thimerosal and autism. However, by 2003, Verstraeten said his study ultimately did not support such a link, and he became a supporter of the vaccines. He was accused of yielding to outside pressure to alter studies indicating a link between thimerosal and autism. One internal CDC document obtained after a FOIA request, showed Verstraeten sent an email that many have interpreted as referring to his difficulty in making the statistical association between thimerosal and autism disappear with the words, "It just won't go away."

In June 2005, *Rolling Stone* published an article written by Robert F. Kennedy Jr. entitled "Deadly Immunity," which claimed that the federal government and the pharmaceutical industry colluded to withhold information concerning vaccine safety. Kennedy also accused Verstraeten of modifying his data to fit the CDC's claim that there is no link between thimerosal and autism, an accusation that Verstraeten has vehemently denied. Yet his personal career choices suggest something sinister: shortly after publishing his findings, Verstraeten left the CDC for a position with the pharmaceutical giant GlaxoSmithKline.

Verstraeten's jump is just another illustration of the "revolving door" policy between government regulators and the corporate world. In 2009, for instance, CDC Director Julie Gerberding left the organization for a job as president of the $5 billion vaccine division of Merck.

Poul Thorsen is another pro-vaccine doctor whose legitimacy has been called into question. Thorson coauthored some of the most frequently cited CDC studies denying the link between thimerosal-containing vaccines and autism. Much of the data cited in these studies remain unavailable to the public. Yet despite the lack of transparency, Thorsen's research has been hailed by the corporate mass media, public health establishment, and Big Pharm as "proof" that there is no connection between vaccines and autism.

In 2014, Thorsen was indicted for fraud and stealing grant money while working for the CDC. The CDC had awarded him grant money for research in Denmark involving infant disabilities, autism, genetic disorders, and fetal alcohol syndrome. According to the U.S. Department of Health and Human Services' inspector general, Thorsen reportedly diverted more than $1 million of the CDC grant money to his own personal bank account and submitted fraudulent invoices on CDC letterhead to medical facilities assisting in the research for reimbursement of work allegedly covered by the grants.

In order to find out who is in the right in the vaccine debate, one need only follow the money. Mass inoculations bring more than $25 billion in revenues to the giant pharmaceutical firms and their hirelings while physicians and researchers who question mass vaccinations make nothing. In fact, many of them risk loss of income and ostracism from the conventional medical establishment. Meanwhile, those who trumpet the benefits of vaccines and downplay their risks can profit enormously.

And the government is also getting in on the act. In 2014, the Obama White House admitted that the CIA used a fake vaccination program in Pakistan to covertly obtain DNA samples from the population as part of the War on Terror. The idea apparently was to use the DNA to locate members of the bin Laden family in Abbottabad, where Osama bin Laden was suspected of hiding. The Pakistani vaccination ruse has raised concerns that U.S.-sponsored health programs will be used to spy on and even infect people. Already, Taliban leaders in Pakistan and Afghanistan have accused legitimate vaccine workers of being spies.

Although White House antiterrorism adviser Lisa Monaco pledged "the agency will not seek to obtain or exploit DNA or other genetic material acquired through such programs," many wonder what's to stop the government from using fake vaccination programs in the future to spread harmful or ineffective substances.

"Oops, that already happened," noted Mike Adams. " It happened right here in the United States, in fact, when 98 million Americans were found to have been injected with polio vaccines contaminated with cancer-causing viruses [Simian virus 40]. In order to make sure no one learned about this deadly vaccine snafu, the CDC quietly removed all accounts of this history from its website."

It's not hard to imagine vaccinations being used for insidious purposes given the regulatory issues attached to them. Vaccination critics are especially concerned about the substantial immunity the federal government gives vaccine makers. The National Childhood Vaccine Injury Act of 1986 created the Vaccine Injury Compensation Program (VICP), which prevents anyone from filing certain civil lawsuits against vaccine manufacturers. Congress handed drug makers this legal protection that in effect gave them a free pass to market as many vaccines as they would like. According to its literature, the VICP is "a no-fault alternative to the traditional tort system for resolving vaccine injury claims that provides compensation to people found to be injured by certain vaccines."

The VICP is funded by a seventy-five-cent excise tax on every vaccine routinely administered to children. This questionable law was upheld by the U.S. Supreme Court in 2011. Although dissenting judge Sonia Sotomayor warned that this decision "leaves a regulatory vacuum in which no one ensures that vaccine manufacturers adequately take account of scientific and technological advancements when designing and distributing their products."

This prompts the question: If inoculations are so safe, why did the federal government need to offer pharmaceutical companies immunity from lawsuits?

Clearly, the cozy relationship between government and pharmaceutical companies qualifies as collusion. And the mainstream media is complicit in this arrangement, refusing to call out vaccine makers even when the evidence is strong. One example of a media leader who spoke up, and paid the price, is Katie Couric, the first female news anchor on a major network. Couric sparked an avalanche of criticism in December 2013 by simply addressing the issue of vaccine safety. Despite the fact that she gave voice to both sides of the debate over Gardasil, the HPV vaccine that has been linked with a number of debilitating side effects, she was savagely attacked as antivaccine.

A vigorous supporter of the HPV vaccine, Alexandra Sifferlin, penned a *Time* editorial charging that Couric was following in the footsteps of Jenny McCarthy, a cohost of the ABC talk show *The View*, who has claimed vaccines bring on autism. "The damage a former Playboy Bunny has been able to do is bad enough. But Couric's misdeeds are all the worse given that she's taken much more seriously than Jenny McCarthy," wrote Sifferlin.

Shortly after Couric's vaccine program aired, she backpedaled in a *Huffington Post* article. "Following the show, and in fact before it even aired, there was criticism that the program was too anti-vaccine and anti-science, and in retrospect, some of that criticism was valid. We simply spent too much time on the serious adverse events that have been reported in very rare cases following the vaccine. More emphasis should have been given to the safety and efficacy of the HPV vaccines," wrote Couric. She added that as a journalist, she felt she could not ignore reports of adverse reactions—including death—to HPV vaccines. "But based on the science, my personal view is that the benefits of the HPV vaccine far outweigh its risks," she added.

"The shaming of Katie Couric for caring and daring to ask questions about Gardasil vaccine, was a well-orchestrated campaign of intimidation. It was a warning delivered to all journalists that—no matter who you are—your character *will* be assassinated if you step out of line and question the safety or effectiveness of a government recommended vaccine," countered Barbara Loe Fisher in a 2014 article for the National Vaccine Information Center.

After noting that the 14,000 annual deaths in the U.S. from six cancers associated with HPV represents less than 3 percent of the more than 550,000 yearly cancer deaths, Fisher concluded, "Whatever the reasons that government officials made HPV vaccination a top public health priority in the U.S., the cyber-lynching of Katie Couric and mothers reporting Gardasil vaccine reactions is a warning to parents everywhere. Do not forget that the cruel, dogmatic position of vaccine risk denialism is: Roll up your sleeve—no questions asked—and 'may the odds be ever in your favor.'"

Often mass media outlets carry lurid stories designed to instill fear of vaccination critics. "An Epidemic of Fear: How Panicked Parents Skipping Shots Endangers Us All" was the headline of an article by Amy Wallace in the October 2009 issue of *Wired*. Wallace detailed the abuse, including death threats, against Dr. Paul Offit, a Philadelphia pediatrician who is co-inventor of a vaccine that some claim can save thousands of lives. Others point to Offit as "the vaccine industry's most well-paid spokesperson."

"This isn't a religious dispute, like the debate over creationism and intelligent design," wrote Wallace. "It's a challenge to traditional science that crosses party, class, and religious lines. It is partly a reaction to Big Pharma's blunders and PR missteps, from Vioxx to illegal marketing ploys, which have encouraged a distrust of experts."

Critics such as Health Ranger Mike Adams have turned the antiscience argument against vaccine supporters. "Note carefully that vaccine zealots are not scientifically-minded people; they are religious zealots who worship the religion of vaccines. Their 'belief' in vaccines is based purely on faith; all evidence be damned! Anyone who studies autism is immediately ostracized and discredited even if their research only hints at a link between autism and vaccines," he wrote. "Anyone who does not conform to the myths and lies of this medical Mafia is subjected to widespread character assassination, where endless lies are spread about them. I've even been accused of being 'anti-science' even though I run a scientific laboratory using atomic spectroscopy equipment to research food safety! In reality, I'm one of the most 'scientific' activists in the country, yet because I express concern of the safety of mercury in vaccines, I too am immediately and viciously branded 'anti-science.' Anyone who is anti-mercury, it turns out, is automatically labeled 'anti-science.'"

It turns out that there are legitimate scientific reasons to be concerned about mercury levels in vaccines. In mid-2014, Adams, working through the Natural News Forensic Food Lab, tested the FluLaval influenza vaccine manufactured by GlaxoSmithKline and found lot 9H2GX contained fifty-one parts per million of mercury, more than twenty-five thousand times higher than the maximum level permitted by the EPA in drinking water. The concentration of mercury in the flu shot was one hundred times greater than the highest level he had ever found in tuna and other ocean fish known for high mercury contamination. "And yet vaccines are *injected* [emphasis in the original] directly into the body, making them many times more toxic than anything ingested orally," he noted.

And despite the toxicity, the efficacy of many vaccines is unclear. By 2005, vaccinations had reduced the annual incidence of mumps in the United States by more than 99 percent. However, the next year a large outbreak occurred among highly vaccinated populations in the United States with similar outbreaks reported worldwide. Eighty-nine percent of those who contracted mumps had already been vaccinated at least twice for the disease, presumably with the controversial measles, mumps, and rubella combination vaccine that has been implicated in causing gastrointestinal disorders and autism. These numbers indicate that the MMR vaccine was, in this case, essentially ineffective in preventing the disease, and offer strong support to parents who would think twice about administering this vaccine to their children.

Despite the risks inherent in vaccines, they won't be going away anytime soon. Vaccine revenues in 2013 reached more than $25 billion, with pharmaceutical profits ensuring government advocacy of the shots. The vaccine market is expected to return a compound annual growth rate of more than 8 percent through 2018. Though vaccines have generated profits for a few massive pharmaceutical firms, the rest of us have suffered, and will continue to do so for the foreseeable future.

CHAPTER 6

GONE TO POT

IF EVER THERE WERE A SINGLE DRUG THAT COULD BE DEFINITELY linked to a variety of social problems—dangerous driving, domestic violence, health issues, and more—it would be alcohol. Yearly, almost eighty-eight thousand Americans die from alcohol-related causes, making it the third leading preventable cause of death in the United States. So it was with noble intentions that prohibitionists fought for decades to have alcohol outlawed, finally succeeding in 1917. It took Americans just over a dozen years to realize that alcohol prohibition was not working and was creating worse problems than alcohol had. Recognizing the failure of this law, Congress passed the Twenty-First Amendment, which officially ended Prohibition.

But when it comes to other drugs, the United States has learned remarkably little from Prohibition, enacting the same failed policies. Ninety years after the sale and use of marijuana was made a crime in the U.S., the so-called War on Drugs continues.

Although a few states in recent years have passed legislation decriminalizing or legalizing marijuana, the majority continue to classify its sale or use as a criminal offense. In 2014, more than 50 percent of federal prison inmates were there for drug offenses, according to the Federal Bureau of Prisons. That percentage has grown over decades from a mere 16 percent in 1970. In Texas, nearly 90 percent of state prison inmates were incarcerated because of drug offenses. This is despite the fact that marijuana, commonly called "pot" or "weed," continues to gain favor with Americans. A 2013 national survey on drug use and health by the Substance Abuse and Mental Health Services Administration (SAMHSA) found marijuana use increased steadily since 2007. The comprehensive survey, involving seventy thousand people above the age of twelve, reported that 7.3 percent of Americans regularly used marijuana in 2012, up from 7 percent in 2011 and significantly higher than the 5.8 percent in the 2007 survey. And self-reporting bias indicates that the true number of Americans who use marijuana may be even higher.

Even those who don't use the drug are increasingly tolerant of its use, as a February 2014 survey from the Pew Research Center confirmed. The survey showed that 67 percent of Americans in all demographic groups thought the government should focus more on providing treatment for, and less on punishing, those who use illegal drugs. Surprisingly, this sentiment cuts across party lines; more than half of those who felt the government should focus more on treatment than prosecution of marijuana identified themselves as Republicans. A full 75 percent of those polled felt the drug will eventually be legalized in all states.

The ACLU advised, "This waning public support [for drug prohibition] is warranted, as evidence continues to document how the War on Drugs has destroyed millions of lives, unfairly impacted communities of color, made drugs cheaper and more potent, caused countless deaths of innocent people caught up in drug war-related armed conflict, and failed to eliminate drug dependence and addiction." In short, Americans are fed up with the current government policy on drugs.

By 2015, twenty-three states and the District of Columbia had legalized marijuana in some form, while four states have legalized it for recreational use, including Alaska and Oregon, where voters in 2014 approved legalization of recreational pot to become effective in 2015. Colorado and Washington previously passed similar ballot measures legalizing marijuana in 2012.

Yet while many Americans support the legalization of marijuana, and a number of states have legalized or decriminalized the drug, in some places the fight against weed continues unabated.

In 1992, there were 812 arrests for small amounts of marijuana in New York City. By 2012, this number had risen to 39,218, and represented almost a million hours wasted by law enforcement officers. Many have pointed to the social costs of draconian marijuana policies. As civil rights organizer Alfredo Carrasquillo put it, "We cannot afford to continue arresting tens of thousands of youth every year for low-level marijuana possession. We can't afford it in terms of the negative effect it has on the future prospects of our youth and we can't afford it in terms of police hours." Such arrests ruin the lives of offenders who otherwise pose no threat to society.

Many well-intentioned people have been unfairly caught up in the antidrug hysteria. In 2013, child protective authorities took the six-month-old daughter of a Michigan couple, Gordon and Maria Green, whom they accused of having marijuana in the home, even though medical marijuana was legal under state law. Critics have accused caseworkers of using scare tactics to intimidate parents, telling them that while their possession of pot is legal under state law, they must surrender their children. Cases such as these go beyond concern over drug use and shed light on the deceptive practices of child protection agencies. In many states, these agencies receive funding based on the number of children they warehouse.

As usual, it is all about the money. According to a 2013 Congressional Research Service report, at least nineteen federal agencies receive billions in antidrug funding today. The nation spends approximately $51 billion yearly on the drug war, with the bulk of this money going to law enforcement. About $24 billion in antidrug money goes to the Department of Justice, the Pentagon, and the Department of Health and Human Services, and even organizations such as the Agriculture Department, the Bureau of Land Management, and the U.S. Forest Service receive significant funds. That makes for a huge number of administrators and staff dependent on the War on Drugs for their financial livelihood. The human cost of these policies is massive. According to the Drug Policy Alliance, there were 1.5 million people in the U.S. arrested on nonviolent drug charges in 2012. Of them, 749,825 were related to marijuana, 658,231 for mere possession.

In an effort to defend their irrational policies, marijuana opponents have grasped at a number of unlikely arguments. In April 2014, Michele Leonhart, the head of the Drug Enforcement Administration (DEA), while testifying to a House Appropriations Subcommittee hearing on the DEA budget, went so far as to attempt to state that pot should remain illegal because of the risk that family pets could become ill from consuming edible marijuana products.

"There was just an article last week, and it was on pets," she told House members. "It was about the unanticipated or unexpected consequences of this [legalization], and how veterinarians now are seeing dogs come in, their pets come in, and being treated because they've been exposed to marijuana."

The article in question appeared in *USA Today*, which stated that the effects of marijuana could make it more difficult for a dog to breathe or vomit up a swallowed item. However, the article also noted that "marijuana itself isn't particularly harmful to dogs" and that dogs are unlikely to eat the drug unless it is baked into a food appealing to them, such as brownies.

Zack Carter, senior political economy reporter for the *Huffington Post*, during a media interview, told how once, while he was in high school, his pet beagle, Pepper, ate an entire ounce of pot contained in some brownies. He said the dog was high for three or four days—falling over, staggering around, and wetting herself—but fully recovered. "I wouldn't want anyone to have that experience although it was pretty funny. The dog ate an entire ounce of weed and was basically fine, so I think the DEA is barking up the wrong tree," he said. Given that a number of common household foods, such as chocolate and pecans, are far more toxic to dogs than marijuana is, using our pets to justify drug prohibition is a questionable line of reasoning at best.

Though the antidrug lobby has largely succeeded in fostering an irrational culture of fear when it comes to drugs, there have been a few legislative steps in the right direction. At the end of May 2014, even pro-pot advocates were taken aback when the House passed an amendment to the appropriations bill that would prevent the DEA and federal prosecutors from targeting medical marijuana in states where it is legal. While pushback from entrenched antipot forces was expected as the appropriations process moved along, passage of the amendment clearly showed a major shift in the thinking of House members, who supported it in a bipartisan vote of 219 to 189.

And in late 2014, Congress, in narrow votes both in the House and Senate, approved the Continuing Resolution Omnibus Bill funding the government that included a de facto declaration of truce in the War on Drugs marijuana front. Included in the legislation to be signed into law by President Obama was a clause forbidding the spending of federal money to prosecute pot possession and sales in states that have approved medical marijuana.

The bill stated, "None of the funds made available in this Act to the Department of Justice [which includes the FBI] may be used, with respect to the States of Alabama, Alaska, Arizona, California, Colorado, Connecticut, Delaware, District of Columbia, Florida, Hawaii, Illinois, Iowa, Kentucky, Maine, Maryland, Massachusetts, Michigan, Minnesota, Mississippi, Missouri, Montana, Nevada, New Hampshire, New Jersey, New Mexico, Oregon, Rhode Island, South Carolina, Tennessee, Utah, Vermont, Washington, and Wisconsin, to prevent such States from implementing their own State laws that authorize the use, distribution, possession, or cultivation of medical marijuana."

Yet overall, the federal government has not been progressive when it comes to drug policy. Instead, the states have taken the lead; the Pew Research Center reported that between 2009 and 2013, forty states moved to ease severe mandatory drug laws.

In August 2013, there was a shift in federal policy. Attorney General Eric Holder announced that the Justice Department would not challenge states that have legalized use of small amounts of marijuana or medical marijuana if the states enacted strict measures to keep the drugs away from minors and took steps to regulate them.

But the actions of the feds belied Holder's words. Even while pledging not to interfere with states that decriminalized pot, federal officers continued to raid pot shops in California. In late October

2013, federal agents attempted to seize the property of California's Berkeley Patients Group (BPG), a medical marijuana dispensary, and essentially shut down the business.

"The Obama administration's ongoing war against patients is despicable and has to stop," said Steph Sherer, executive director of Americans for Safe Access. "This is a mean, vindictive move aimed at shutting down one of the oldest and most well-respected dispensaries in the country." Berkeley City Council member Darryl Moore agreed, offering a resolution that stated, "BPG has served as a national model of the not-for-profit, services-based medical cannabis dispensary."

Moore claimed such dispensaries improve lives and assist in end-of-life transitions of thousands of patients and have been significant donors to dozens of other organizations which have shaped local, state, and national policies around medical cannabis.

Even as voters indicate a willingness to soften pot laws, federal machinery works against the marijuana business.

In 1982, Congress amended the U.S. tax code to include section 280E, which prevents businesses selling a Schedule I or II drug, such as heroin, methamphetamine, cocaine, and marijuana, from deducting all normal business expenses. This code was enacted as the Supreme Court had ruled that even illegal businesses must pay taxes but cannot claim the usual deductions.

Despite pleas to amend the tax laws, an IRS spokesperson said only Congress can amend the Internal Revenue Code or the Controlled Substance Act.

"I believe that the feds extend the drug war through 280E," said Jordan Cornelius, a Denver accountant who has worked with marijuana companies in Colorado. "If [the federal government] can't put them out of business legally when voters are mandating these businesses to move forward, it's very easy to put them out of business financially . . . A lot of times, instead of paying a tax rate that should be 30 to 40 percent, they are paying rates between 80 or 90 percent," Cornelius explained. "I even have a client right now that is paying more than 100 percent effective tax rate."

"The problem is that we have passed laws that allowed these

medical marijuana and recreational marijuana companies to do business," said University of Denver finance professor Mac Clouse. "But we have all these other laws, tax laws, federal laws that make it incredibly difficult if not utterly impossible to survive."

Under the oversight of the federal government, the DEA has been aggressive in pursuing leads connected to the War on Drugs. In testimony to the House, DEA Chief Michele Leonhart was asked if DEA agents were demoralized by the rising tide of pro-pot legislation and the decision of the Justice Department not to challenge the states' decisions on marijuana. "Actually, it makes us fight harder," she responded, reflecting the division in opinions within the government.

Perhaps when someone determines how to corner the market on medical marijuana for some large corporation, the feds will finally relax their antipot campaign. Indeed, one reason the federal government may be unwilling to legalize marijuana is that they've yet to figure out how to make money off of it.

Some suspect the federal government's crackdown on pot is merely an attempt to squelch any competition to its own medical marijuana business. On October 7, 2003, a patent (number 6,630,507) for medical marijuana was granted and assigned to "The United States of America as represented by the Department of Health and Human Services (Washington, DC)."

Titled "Cannabinoids as Antioxidants and Neuroprotectants," the patent abstract claimed "cannabinoids [the chemicals found in cannabis or marijuana] have been found to have antioxidant properties . . . This newfound property makes cannabinoids useful in the treatment and prophylaxis of wide variety of oxidation associated diseases, such as ischemic, age-related, inflammatory and autoimmune diseases."

The U.S. government may be able to profit off of marijuana in ways that extend past recreational use. This patent for medical marijuana comes at a time when studies are finding new medical uses for the psychoactive properties of pot. Research at the Skaggs Institute for Chemical Biology found the cannabinoid molecules in tetrahydrocannabinol (THC) could slow or even stop the progress of Alzheimer's disease. "Compared to currently approved drugs prescribed for the treatment of Alzheimer's disease, THC is a considerably superior inhibitor of [peptide] aggregation, and this study provides a previously unrecognized molecular mechanism through which cannabinoid molecules may directly impact the progression of this debilitating disease," said study author Lisa M. Eubanks. The promise THC holds as a potential curative for Alzheimer's has even been acknowledged by the National Institutes of Health.

Sayer Ji, writing in *Waking Times*, argues that if the medical benefits of cannabis found in peer-reviewed studies were made public, they would include one hundred "proven therapeutic actions of this amazing plant." These would include: analgesic (pain-killing), antispasmodic, anti-inflammatory, antidepressive, bronchodilator, antioxidant, cardioprotective, antitumor, enzyme inhibitor, antipsychotic, and many others. Thanks to scientific investigation, the idea that marijuana has beneficial medicinal qualities is no longer merely theoretical.

The elitist-controlled medical community, in league with corporate Big Pharma, has created a medical-industrial complex that does not want consumers using natural compounds for healing. Such use of naturally grown remedies might threaten the medical business model, which may explain the aggressive stance of both doctors and government toward homeopathic treatments. It also may explain why the prison population in America consisting of more than six million people, exceeds the number of prisoners held in the gulags of the former Soviet Union at any point in its history.

The production of hemp is another area where the DEA has been overly aggressive in pursuing potential offenders. Congress has become more lenient on hemp, and the 2014 Farm Bill eased restrictions on growing it. The bill also contained a clause allowing colleges and universities to grow hemp for research purposes. On the state level, more than a dozen states, including Oregon, Montana, Colorado, North Dakota, California, Kentucky, Vermont, West Virginia, and Maine, have legalized hemp, used for paper, clothing, rope, and other practical applications.

Michael Boldin, executive director of the Tenth Amendment

Center, said that while production is not completely legal, many states are growing hemp. "Fields of hemp are growing in Colorado already," noted Boldin in early 2014. "Vermont legalized late last summer, and multiple other states are considering bills to do the same this year. Credit doesn't go to the feds on this one. It belongs to the states, which had the courage to stand up to unconstitutional laws, and force the change on a national level."

Yet the DEA has obstructed local efforts to grow hemp. In May 2014, Kentucky agriculture commissioner James Comer sought an injunction to stop federal officials from halting the importation of 286 pounds of Italian hemp seeds ordered by the department, which needed to be planted by the beginning of June. The plans of local farmers who intended to plant the seeds and harvest the hemp were stalled until U.S. District Judge John G. Heyburn II made a decision. The DEA had ordered U.S. Customs to hold the seeds because they said Kentucky officials failed to get a controlled-substance import permit.

This sort of interference rightfully drew the ire of a number of congressmen. Senate minority leader Mitch McConnell called on the DEA to release the hemp seed, noting the federal Farm Bill passed earlier in the year allowed state departments of agriculture, along with universities, to grow industrial hemp for research purposes. In a statement, McConnell remarked, "It is an outrage that DEA is using finite taxpayer dollars to impound legal industrial hemp seeds. The agency should immediately release the hemp seeds so Kentucky pilot projects can get under way, which will ultimately lead to more economic opportunities in our state." Senator Rand Paul echoed McConnell, saying the federal impounding of the seeds was "completely unacceptable."

The politicians' support, along with public pressure, worked. On May 23, the DEA approved a permit and released the seeds for research. Adam Watson, the industrial hemp coordinator for the state agriculture department, reiterated the essentially harmless nature of the project, remarking, "It's been such a long period of time since any hemp has been grown in Kentucky [that] we really don't know what we're dealing with. So the purpose of all these research projects is to evaluate how these varieties perform and what's the best method for Kentucky producers to grow it." Although hemp and marijuana are of the same species, *Cannabis sativa*, hemp has a lesser amount of the euphoria-producing psychoactive compound tetrahydrocannabinol (THC) than marijuana. One wonders why the DEA is so focused on crackdowns on safe products such as hemp when they could be turning their efforts to more dangerous illicit drugs.

Federal obstructionism continued in 2014 when the U.S. Bureau of Reclamation announced that they would continue to prevent pot growers from using federal irrigation waters. Since 1902, the bureau has been charged with maintaining dams, power plants, and canals in seventeen western states, including Washington and California, two states that have legalized recreational marijuana use.

"As a federal agency, Reclamation is obligated to adhere to federal law in the conduct of its responsibilities to the American people," stated Dan DuBray, the bureau's chief of public affairs. Ellen Canale, a spokeswoman for Holder's Justice Department chimed in, saying, "The Department of Justice will continue to enforce the Controlled Substances Act and will focus federal resources on the most significant threats to our communities." The cultivation of marijuana is apparently considered to be a significant threat.

The DOJ outlined eight factors in particular on which it would focus, most of them not presenting any real problem. They include: preventing the distribution of marijuana to minors, a task already reasonably met with alcohol and tobacco; preventing revenue from the sale of the drug from going to criminal enterprises, which seems no different from money going to illegal activities through legal businesses; preventing diversion to states where it is illegal, this assuming it will remain illegal; preventing state-authorized marijuana activity from being used as a cover for other illegal drug activity, as if this is not already happening; preventing violence and the use of firearms in the cultivation of marijuana, which could be accomplished by legalization; preventing drugged driving and other adverse public health consequences, as if there are not DUI laws now; preventing the cultivation of marijuana on public lands; and preventing marijuana possession or use on federal property, as if this is not already happening.

Growers and others connected to the marijuana industry are resigned to interference from the feds. "We're used to this kind of treatment," said Elan Nelson, a business consultant for Medicine Man dispensary in Denver. "The federal government is looking for one obstacle after another to place hurdles before this industry," Nelson warned. "Pretty soon it's going to be air. They're going to say you can't use the air because it belongs to the federal government. It's just ridiculous."

The feds' opposition to marijuana cultivation has even impacted the banking community. Colorado's two largest banks, Wells Fargo and FirstBank, have attempted to cut off the pot industry, refusing to offer new loans to landowners with preexisting leases for pot businesses. Wells Fargo also gave commercial real estate clients an ultimatum: either evict marijuana businesses or seek refinancing elsewhere. Wells Fargo defends its policy by pointing out that such loans theoretically are subject to federal drug-seizure laws, putting the bank at risk. "Our policy of not banking marijuana-related businesses and not lending on commercial properties leased by marijuana-related businesses is based on applicable federal laws," explained Wells Fargo spokeswoman Cristie Drumm. Again, the banking industry is at the mercy of the federal government in its crusade against pot.

Similar banking restrictions were the subject of a 2014 lawsuit filed by medical marijuana dispensary Allgreens LLC against the IRS. The IRS does not accept tax payments in cash, putting dispensaries, which are often unable to open bank accounts, in a bind, as Allgreens pointed out. "The taxpayer is unable to secure a bank account due to the nature of its business. With no bank account and no access to banking services, the taxpayer is simply incapable of making the payments electronically," explained Allgreens' attorney Rachel Gillette. She added that the IRS suggested the firm pay a third party to deliver the taxes, a suggestion that seems to amount to tacit IRS approval of money laundering.

The threat of selective federal enforcement was accentuated in 2013, when Obama's drug czar, Gil Kerlikowske, repeatedly stated the administration had a "zero tolerance policy" toward drug use, and federal efforts to punish pot smokers reached new heights of absurdity.

In October 2013, Angela Kirking, an Illinois face-painting artist, was awakened in her home by agents of the Drug Enforcement Administration (DEA) with guns drawn. "They were in full attack mode, came at me guns raised, flashlights. Just like you see in the movies," Kirking said. Her crime? Three weeks earlier she had purchased a bottle of organic fertilizer from a nearby garden store being monitored by the DEA for persons buying hydroponic equipment, which could be used to grow pot. The raid, consisting of four DEA agents and five Shorewood, Illinois, police officers, included a thorough search of Kirking's household trash, where one officer claimed he detected "a strong odor of green cannabis." Further searching revealed 9.3 grams of pot, a small quantity for personal use, in the woman's art room.

The raid raises troubling questions about the threshold of evidence required for law enforcement to obtain a warrant to search private property. The "odor" angle may see increased use by police as more states legalize pot. Denver, for example, recently passed an odor ordinance with a hefty fine of up to $2,000 for polluting the air. In an effort to settle arguments over whether smoking pot in one's own home violates the law, the city defined smoking as an offense when the smoke is mixed with seven times the volume of clean air, creating a strong odor. Ben Siller, a member of Denver's Department of Environmental Health, moves about town with an olfactometer, a device to check if the smell of marijuana is breaking the odor law. Many feel the law itself stinks.

A similar raid occurred in the Kansas City, Kansas, suburb of Leawood, where lawmen with bulletproof vests and using SWAT team attack tactics raided the home of Robert and Adlynn Harte after Robert had shopped at a hydroponic garden store for materials to build an indoor vegetable garden for his son. The Hartes, it turned out, were former CIA employees and no drugs were found, although one officer suggested their thirteen-year-old son might have used pot.

These incidents are just two examples of the activist drug policing carried out across the country. The ultimate effect of federal government overreach on marijuana policy, of course, is seen in our prisons. Pot prohibition in the U.S. has resulted in wrongful arrests and ruined lives for offenders, not to mention overcrowded prisons mostly filled with drug offenders.

An ACLU report released in June 2013 showed that while marijuana arrests have decreased since 2006, there are still significantly more of them than there were in the early 2000s. Nearly half of all drug arrests are for pot, and often for minor transgressions: a full 88 percent of marijuana arrests are for simple possession. The report also indicated a racial bias in pot busts, as blacks are 3.73 times more likely to be arrested for pot possession than whites.

The report also noted the financial cost of our obsession with drug prohibition. Despite the fact that many states face pension shortfalls and budget crunches, marijuana-related arrests continue to drain public coffers. The ACLU estimated that nationally, the cost of arrests just for possessing marijuana runs somewhere between $1 billion and $6 billion per year. The states spending the most money per capita arresting pot users were New York (which spends almost twice as much as any other state), Maryland, Illinois, Wyoming, Nevada, Delaware, New Jersey, Connecticut, and Arizona.

And many states waste significant funds keeping prisoners incarcerated, not to mention the human costs of these draconian drug regulations. In Texas, for instance, nineteen-year-old Jacob Lavoro faced a life sentence in prison for baking and selling brownies laced with pot. Lavoro's crime was listed as a first-degree felony because his brownie recipe included hash oil, which allowed prosecutors to include the sugar, cocoa, butter, and other ingredients when determining the weight of the drugs. His father, Joe Lavoro, complained, "Five years to life? I'm sorry. I'm a law-abiding citizen. I'm a conservative. I love my country. I'm a Vietnam veteran, but I'll be damned. This is wrong." Even his son's attorney, Jack Holmes, questioned the law, stating, "I was outraged. I've been doing this twenty-two years as a lawyer and I've got ten years as a police officer and I've never seen anything like this before."

Such stiff penalties are in part the legacy of the Controlled Substances Act of 1970. In this piece of legislation, the DEA classified pot as a Schedule I drug, placing it in the same category as far more dangerous drugs such as heroin. Other drugs classified as Schedule I include cocaine, crack cocaine, amphetamines, and crystal meth; meanwhile alcohol and tobacco, which are responsible for the greatest number of drug-related deaths in the United States, are not scheduled at all.

"We're stuck in a catch-22," said Tamar Todd, senior staff attorney for the Drug Policy Alliance. "The DEA is saying that marijuana needs FDA approval to be removed from Schedule I, but at the same time they are obstructing that very research. While there is a plethora of scientific evidence establishing marijuana's safety and efficacy, the specific clinical trials necessary to gain FDA approval have long been obstructed by the federal government itself.

"The scheduling is made within the context of a law enforcement agency and that law enforcement agency has an interest in keeping drugs illegal and maintaining the status quo," she added.

Adding the federal restrictions on marijuana research to the IRS sanctions against pot business deductions and to the DOJ's continued enforcement of old drug laws, it is apparent that those who control the federal government machinery do not want the public to have easy access to pot.

Another federal catch-22 involves the Bureau of Alcohol, Tobacco, Firearms and Explosives (BATF), which on its Form 4473 asks potential firearm purchasers: "Are you an unlawful user of, or addicted to, marijuana or any depressant, stimulant, narcotic drug, or other controlled substance?" Even if your state has legalized pot, if you are a user, you must reply yes, which then opens you to search under "reasonable cause" as a user of a controlled substance. Such persons are prohibited by federal law from their constitutional right to possess, trade in, or transfer firearms or ammunition. All this despite the fact that marijuana is one of the few drugs for which there is no lethal dose and no proven long-term harm. On the other hand, tobacco causes forty-six times more deaths than all illegal drugs combined, while prescription drugs cause twice as many deaths as all illegal drugs combined.

Furthermore, harsh policies on marijuana do not lead to lower rates of drug abuse. A recent World Health Organization survey showed that the United States has the highest level of illegal drug use in the world. Meanwhile, forward-thinking countries are liberalizing their drug policies, following the lead of Uruguay, which in 2014 became the first national government to grow, regulate, and safely distribute pot.

Although many U.S. lawmakers remain stubbornly resistant to such a move, some are beginning to come around, thanks to the oil that has always greased the American political system: money. States that have legalized marijuana have already reaped tremendous financial gains. As of February 2014, the state of Colorado had already collected at least $100 million in revenue from marijuana taxes. Colorado governor John Hickenlooper predicted that 2015 sales and excise taxes on marijuana would generate $98 million for the state, well above the $70 million annual estimate floated when voters approved the taxes. In Washington, budget forecasters predicted the state's new legal recreational marijuana market could swell coffers by more than $190 million over four years starting in mid-2015, a fact that has not been lost on many observers. "Voters and state lawmakers around the country are watching how this system unfolds in Colorado [and Washington], and the prospect of generating significant revenue while eliminating the underground marijuana market is increasingly appealing," said Mason Tvert, a legalization activist for the Marijuana Policy Project. Should other states also decide to legalize the drug and tax it aggressively, they would stand to gain similar benefits.

Beyond tax revenues, the legalization drive has also benefited the economy by creating a whole new industry with new jobs and opportunities. By mid-2014, Colorado had established a huge business in pot, with about 340 medicinal and recreational pot shops open in Denver alone. More impressively, fears over the legalization of weed appeared unfounded, with both crime and traffic fatalities down. According to Denver police, burglaries and robberies dropped between 4 and 5 percent in the first four months of 2014. Contrary to some expectations, data released by the Colorado Department of Public Health and Environment in mid-2014 showed pot use among high school students had stagnated, and was actually beginning to decline. The percentage of teens who reported using marijuana in the previous month dropped from 22 percent in 2011 to 20 percent in 2013.

Traffic fatalities also decreased following legalization, to the surprise of some observers. Yet research has suggested a link between liberal marijuana policies and low traffic fatality rates. Studies from the Dutch Institute for Road Safety, the U.S. National Transportation Safety Administration, the United Kingdom Transport Research Lab, Colorado University, and Montana State University have all come to the same conclusion: postlegalization, Colorado experienced a drop in both traffic fatalities and beer sales. The studies indicate that there is crossover between recreational drinkers and marijuana users, and that marijuana users are less likely to take risks and drive recklessly.

It should be noted, however, that those supporting decriminalization of pot do not necessarily support recreational use of the drug.

Natural health activist Mike Adams explained it this way: "I am wholeheartedly in support of marijuana decriminalization, yet at the same time I strongly encourage people to avoid smoking it. That's the difference between me and a police state government, by the way—I believe in your right to choose what you wish to do with your body, while the police state government would far prefer to shove a gun in your face, slap a pair of handcuffs on your wrists and throw you into the prison system, which is little more than a modern-day slave labor camp that benefits corporate interests under the guise of fighting the 'war on drugs.'"

As with any other substance capable of abuse, there are many legitimate reasons to control the sale and use of marijuana, just as with alcohol and tobacco. Many people have trouble distinguishing between the decriminalization of pot, which would save lives, careers, and lessen the prison population, and its total legalization, including recreational use.

With so many positive consequences of the legalization of marijuana in Colorado, it's fair to ask why so many remain opposed to reforming drug policy. Again, it all comes back to the money. A number of pot opponents, particularly those with impressive credentials, may be under the influence of Big Pharma, as Lee Fang suggested in a 2014 article in *Vice* magazine. "*Vice* has found that many of the researchers who have advocated against legalizing pot have also been on the payroll of leading pharmaceutical firms with products that could be easily replaced by using marijuana," wrote Fang. "When these individuals have been quoted in the media, their drug-industry ties have not been revealed."

One such expert, Dr. Herbert Kleber of Columbia University, is vocal about the dangers of marijuana and has been quoted in such media outlets as CBS, NPR, and CNBC. Kleber's warnings against marijuana have been cited by the New York State Association of Chiefs of Police and the American Psychiatric Association. "What's left unsaid is that Kleber has served as a paid consultant to leading prescription drug companies, including Purdue Pharma (the maker of OxyContin), Reckitt Benckiser (the producer of a painkiller called Nurofen), and Alkermes (the producer of a powerful new opioid called Zohydro)," notes Fang.

Other academic opponents to marijuana legalization cited by Fang include associate professor of psychiatry at Harvard Medical School Dr. A. Eden Evins, a board member of the antipot organization Project SAM. Following Evins's collaboration in an article on marijuana legalization for the *Journal of Clinical Psychiatry,* the publication found that as of November 2012 she was a consultant for Pfizer and DLA Piper, a law firm which in 2014 advised Pfizer on the $635 million acquisition of Baxter commercial vaccines. She also had received grant/research support from Envivo, GlaxoSmithKline, as well as Pfizer,

Another outspoken pot critic was Dr. Mark L. Kraus, a board member of the American Society of Addiction Medicine. Dr. Kraus opposed a proposed medical marijuana law offered in 2012 in Connecticut. According to Fang's research, financial disclosures showed Kraus had served on a scientific advisory panel for painkiller producers such as Pfizer and Reckitt Benckiser in the year prior to his activism against the bill. Fang said none of the experts named responded to his requests for comment. Such conflicts of interest further help to explain the reasons for the crusade against marijuana.

Our government policy on pot has been misguided since the

original fears of "reefer madness," which date to the first half of the twentieth century. Like vaccines, our attitude toward illicit drugs has been corrupted and secretly guided by a small number of elites who profit from prohibition. In cases where marijuana has been legalized, the citizens have benefited, but powerful, entrenched forces have helped to stem the tide of legalization. The situation is reprehensible. Yet even more disconcerting is the damage being done by entirely legal prescription drugs.

CHAPTER 7

PSYCHIATRIC DRUGS AND SHOOTERS

DURING THE DEBATE OVER MARIJUANA LEGALIZATION, A DARK side to the use of legal drugs was largely overlooked. Because of the overuse and mishandling of legal prescription drugs, America has experienced a terrifying rise in teen suicides, school shootings, and the deaths of many veterans. Then secretary of veteran affairs Eric Shinseki reported in November 2009 that "more veterans have committed suicide since 2001 than we have lost on the battlefields of Iraq and Afghanistan."

In May 2014, another shooting by a seemingly average kid rocked the nation. Twenty-two-year-old Elliot Rodger killed six people and injured thirteen others in a rampage in Isla Vista, California, that ended with his taking his own life.

Immediately, his father and the corporate mass media began the drumbeat, blaming guns for the carnage despite the fact that Rodger's first three kills were with a knife. There were no calls for registering or outlawing bladed instruments. Although Rodger also rammed at least one victim with his car, no one called for outlawing automobiles. In a "manifesto" left behind, Rodger stated, "I was different because I am of mixed race. I am half White, half Asian, and this made me different from the normal fully-white kids that I was trying to fit in with. I envied the cool kids, and I wanted to be one of them." This prompted cries that racial hatred was to blame for Rodger's crime.

One of Rodger's victims, twenty-year-old Bianca de Kock, who survived five gunshots, told how he wore a "smirky, grimacy smile" as he gunned down her sorority sisters. "He wanted to do this, he looked happy about it," she told the news media.

Elliot Rodger was only the latest in a string of young men involved in shooting incidents. The corporate mass media always blames firearms for the shootings. But a look at history reveals that prior to the 1968 Gun Control Act, any person could legally own all sorts of weapons, including machine guns. Yet these guns prompted no school shootings.

What the media fails to mention is the common denominator in so many of these shootings: psychiatric drugs. Of course, gunmakers, unlike pharmaceutical companies, do not spend millions on advertising. Few in the media have detailed the fact that Rodger was being treated for psychological and psychiatric issues. In his own words, he said, "I will quickly swallow all of the Xanax and Vicodin pills I have left . . ." Is it not possible that Rodger's mental illness, and the psychiatric drugs with which he was being treated, are at least partially responsible for his heinous deeds?

As Rodger's story illustrates, drugs may play a significant role in the recent spate of senseless public shootings, though in this case the drugs are legal. And as we see from Rodger's case, these drugs are being prescribed to more and more young people, frequently under the age of eighteen. When dangerous psychiatric medications are mixed with the typical tensions of adolescent life, the results can be deadly.

Rather than continually send heartfelt condolences to the families of the victims, it is time for lawmakers to investigate the connection between prescription psychiatric drugs and violence. In nearly every school shooting, including the 1999 tragedy at Columbine High School, the shooters were medicated.

In July 2012, James Holmes walked into a midnight showing

of a Batman movie in Aurora, Colorado, and killed twelve people, wounding fifty-eight others. The *Denver Post* reported Holmes was taking a generic form of the drug Zoloft. The fact that Seung-Hui Cho, who killed thirty-two persons at Virginia Tech in April 2007, had been undergoing psychological counseling involving the use of prescription psychoactive drugs was rarely mentioned in the mass media. In fact, the prescription drug use of other mass murderers remains unclear because authorities have sealed their medical records from the public. It is known that Army Specialist Ivan Lopez, who killed four persons including himself in an April 2014 shooting spree at Fort Hood, Texas, was being treated with Ambien and other medication for anxiety and depression. Likewise, Aaron Alexis, who killed twelve persons at the Washington Navy Yard in 2013, was being treated with the antidepressant Trazadone.

Sometimes the psychiatric drugs themselves are a factor as withdrawal from selective seratonin reuptake inhibitors (SSRIs) can be particularly unpleasant. British psychiatrist Dr. David Healy notes that "almost all the school shooters that we know of have either been on or using these drugs or in withdrawal from them," a condition called SSRI discontinuation syndrome. While certain states of depression undoubtedly can be soothed by such SSRIs as Paxil, Prozac, Zoloft, Effexor, and others, Dr. Healy warned, "You can become emotionally numb when you go on these drugs. That means you can do things you wouldn't normally contemplate doing."

The website SSRI Stories.org has begun to explore the potentially deadly effects of these drugs. It offers a collection of more than five thousand accounts from popular media and scientific journals in which prescription drugs were associated with a variety of deviant acts, many of them violent. "Withdrawal, especially abrupt withdrawal, from any of these medications can cause severe neuropsychiatric and physical symptoms," warned a post on this site. "Withdrawal is sometimes more severe than the original symptoms or problems."

Several stories recounted on the site involved the violent behavior of patients who had gone off their medication. One example was the tragedy of thirty-five-year-old Derek Ward, who in October 2014 decapitated his mother, Pat, who had scheduled an appointment with a psychiatrist in two days to replenish her son's SSRI medications. Ward, who according to his family had been off his medications only four days, later died when he threw himself in front of a train.

Even the government has acknowledged the dangers of SSRI drugs. In 2004, the FDA ordered pharmacies to provide to all parents or guardians for youngsters eighteen and under an "Antidepressant Patient Medication Guide," which stated in part, "Call healthcare provider right away if you or your family member has any of the following symptoms: Acting aggressive, being angry, or violent and acting on dangerous impulses . . . Never stop an antidepressant medicine without first talking to a healthcare provider. Stopping an antidepressant medicine suddenly can cause other symptoms."

There are about twenty-five million Americans on SSRI drugs. If only one-tenth of one percent react violently to the drugs, that's still twenty-five thousand potential mass murderers.

Of greatest concern is the frequency with which these drugs are being prescribed. Early on, only severe cases of depression or anxiety were thought to warrant psychiatric drugs. But today virtually any school kid may be subject to drug treatment, sometimes even against their will or their parents' knowledge.

One such example is Chelsea Rhoades, a 15-year-old sophomore at Penn High School in Mishawaka, Indiana, who was pulled out of her classroom and told to sign a form, then was administered a TeenScreen psychological assessment test. The test included such questions as: "Have you had trouble sleeping, that is, trouble falling asleep, staying asleep, or waking up too early? Have you had less energy than you usually do? Has doing even little things made you feel really tired? Has it often been hard for you to make up your mind or to make decisions? Have you often had trouble keeping your mind on your schoolwork/work or other things? Have you often felt grouchy or irritable and often in a bad mood, when even little things would make you mad? Have you gained a lot of weight, more than just a few pounds? Have you lost weight, more than just a few pounds?"

Like many teen students, Chelsea did as she was told and never

asked if the test was voluntary. After the test, she was told she suffered from "Obsessive Compulsive Disorder for cleaning and social
anxiety disorder." She was told to seek medical treatment.

Dr. Julian Whitaker, author and founder of the Whitaker Wellness Institute in Newport Beach, California, asked, "Since when are
issues with sleeping, energy, and feeling tired indicative of a mental
illness? What kid—or adult for that matter—hasn't at one time or
another in the last four weeks felt indecisive, unfocused, grouchy, or
irritable? And the questions about weight are just plain nuts. Adoles-
cents are expected to have growth spurts!"

But other test questions were less anodyne. They included "Have
you thought seriously about killing yourself? Have you tried to kill
yourself in the last four weeks? Have you ever in your whole life
tried to kill yourself or made a suicide attempt?" As Dr. Whitaker
remarked, "One thing [is] for certain, if they'd never given [any]
thought [to] suicide, they will now."

The girl's parents, Michael and Teresa Rhoades, filed suit against
the school claiming their parental rights were violated since the
school administered the psychological test without their permission,
did not clearly state the test was voluntary, and made a diagnosis
based on Chelsea's test answers. Representatives of the school, arguing the TeenScreen test was optional and confidential, moved for
a summary judgment from U.S. District Court for the Northern
District of Indiana, expecting to win as they usually do. However,
noting that the school had not made clear that the test was voluntary and had not even filled in the name of any person or persons
to whom the questions were to be referred, the court ruled in the
parents' favor. This particular case was a victory for parents, privacy,
and common sense, but many other children in U.S. schools have
not been so lucky.

A 2014 study based on information from the CDC's National
Center for Health Statistics indicated that almost 11 percent of schoolchildren ages four to seventeen are now taking psychiatric drugs for
emotional or behavioral problems. Many of these medications pur-
port to treat attention deficit hyperactivity disorder (ADHD), a condition in the past described simply as restlessness or acting out of

boredom. Such medications disproportionately harm certain groups of children. The study found that children from poorer families were more likely to be medicated than their well-to-do peers, and that boys were more frequently prescribed psychiatric meds than girls.

Yet the science behind the so-called ADHD epidemic is not well supported. Some researchers, such as behavioral neurologist Richard Saul, do not even believe that ADHD exists. Noting that the number of adults taking a drug for ADHD has increased 53 percent between 2008 and 2012 and nearly doubled for young persons, Saul noted, "Today, the fifth edition of the DSM only requires one to exhibit five of eighteen possible symptoms to qualify for an ADHD diagnosis. If you haven't seen the list, look it up. It will probably bother you. How many of us can claim that we have difficulty with organization or a tendency to lose things; that we are frequently forgetful or distracted or fail to pay close attention to details? Under these subjective criteria, the entire U.S. population could potentially qualify. We've all had these moments, and in moderate amounts they're a normal part of the human condition."

As more than twenty conditions can lead to symptoms similar to ADHD, attention deficient hyperactivity disorder has too often become a diagnosis of choice that saves time for doctors and generates huge profits for the drug corporations.

Even more disturbing is the number of very young children who are being prescribed these medications. According to a report by the CDC, more than ten thousand American toddlers on Medicaid were being medicated for maladies outside established pediatric guidelines, and an additional four thousand children covered under private insurance plans were being medicated.

"It's absolutely shocking, and it shouldn't be happening," said a children's mental health consultant to the Carter Center. "People are just feeling around in the dark. We obviously don't have our act together for little children," commented Anita Zervigon-Hakes, a consultant to the Carter Center in Atlanta, where the report on children was presented in May 2014.

Dr. Susanna N. Visser, who oversaw the CDC's ADHD research, agreed. "Families of toddlers with behavioral problems are coming to the doctor's office for help, and the help they're getting too often is a prescription for a Class II controlled substance, which has not been established as safe for that young of a child. It puts these children and their developing minds at risk, and their health is at risk," warned Dr. Visser.

Dr. Nancy Rappaport, a child psychiatrist and director of school-based programs at the Cambridge Health Alliance outside Boston, specializes in underprivileged youth, many coming from broken homes. "In acting out and being hard to control, they're signaling the chaos in their environment," she remarked. "Of course only some homes are like this—but if you have a family with domestic violence, drug or alcohol abuse or a parent neglecting a two-year-old, the kid might look impulsive or aggressive. And the parent might just want a quick fix, and the easiest thing to do is medicate. It's a travesty."

Where once a daydreaming child was simply chastised by a teacher who ordered them to stay with the rest of the class, today they are sent to the school nurse, who, more often than not, refers them to a psychologist inclined to prescribe medications such as Ritalin and Adderall. These drugs can calm a child's hyperactivity, but also have their share of serious potential side effects, including growth suppression, insomnia, and hallucinations.

It's easy to see why it might not be the best idea to cause young children to experience these effects. The American Academy of Pediatrics guidelines do not cover children below the age of four, because the academy considers hyperactivity developmentally appropriate for toddlers and understands that more time is needed to see if a disorder is truly present.

Despite this policy, one of the worst offenders in this unprecedented drugging of youth is Medicaid. As reported by the Alliance for Natural Health (ANH) in 2013, the number of children—many of them under the age of three—on Medicaid who are taking antipsychotic drugs has tripled in just ten years. Between 1999 and 2008, the amount Medicaid spends on antipsychotic drugs more than doubled. The program currently spends $3.8 billion annually on antipsychotics, more than it spends on any other class of drugs.

A spokesperson for the Texas Health and Human Services Commission defended such drugging by claiming the prescriptions were to help infants "with discomfort." Many of the antipsychotics used today are second- and third-generation antipsychotics, with bizarre names such as Abilify (the nation's top-selling prescription drug), Risperdal, Seroquel, and Zyprexa. These drugs have replaced first-generation antipsychotics such as Haldol and Thorazine, though the side effects remain. As the ANH notes, "Keep in mind that common side effects for one common psychiatric medication include a shuffling walk, drooling, rapid weight gain, and the inability to speak." This list would give any adult pause; the fact that these drugs are being prescribed to infants is unconscionable.

And the financial repercussions are equally damning. Due to the large number of people receiving antipsychotics who are on Medicaid, a full 70 percent of the total spent on these drugs in the U.S. is footed by the taxpayers. "This is nothing less than crony capitalism in action," noted the ANH. "Pharmaceutical companies donate heavily to political campaigns, and legislators pass laws that compensate them in spades for the amount donated. Those who pay the real price are the poor and powerless." The organization added that there is no corporate incentive for such essentials as nutrition, exercise, and love, as these pursuits cannot be patented.

In addition to the mind-warping aspects of psychiatric drugs, recent studies show that memories can now be both created and erased through the use of drugs and the use of light to stimulate certain sections of the brain. Neuroscientists at the University of California, San Diego, found that by using light to activate neurons in the brains of rats, they were able to activate certain proteins that induced false memories. Research team leader Roberto Malinow reported, "We can make a memory of something that the animal never experienced before." He added, "We were playing with memory like a yo-yo," If this can be done in rats, it can be done in humans.

Some researchers believe it is entirely possible that the globalists even now are attempting to manipulate the minds of American citizens by funding research such as that in San Diego as well as supporting legislation for mass mental analysis. The U.S. Preventative

Services Task Force has urged routine screening for all American teenagers for depression, and politicians were ready to step up to the plate. Congress has periodically introduced legislation proposing widespread mental health screening.

In 2007, legislation was introduced entitled the Postpartum Mood Disorders Prevention Act, which called for the mental screening of mothers for signs of depression. Such mandatory screening for depression may soon become state law in Illinois, and similar legislation has already been adopted or at least introduced in several other states. In 2009, this mass screening scheme was proposed again in the form of the Melanie Blocker-Stokes Mom's Opportunity to Access Health, Education, Research, and Support for Postpartum Depression Act of 2009, otherwise known simply as the Mother's Act. This law was reintroduced into both bodies of the new Congress in January 2009, after the 2007 bill died in the Senate.

Critics of the Mother's Act, fearful that the legislation would mandate even further drugging of both mothers and infants, were incensed to learn that the same provisions were contained within President Obama's 2010 Affordable Care Act.

In an article entitled "Branding Pregnancy as a Mental Illness," Byron Richards noted, "The Mother's Act has the net effect of reclassifying the natural process of pregnancy and birth as a mental disorder that requires the use of unproven and extremely dangerous psychotropic medications (which can also easily harm the child). The bill was obviously written by the Big Pharma lobby and its passage into law would be considered laughable except that it is actually happening."

Investigative journalist Evelyn Pringle, writing for the political newsletter *Counterpunch*, wrote, "The true goal of the promoters of this act is to transform women of child bearing age into life-long consumers of psychiatric treatment by screening women for a whole list of 'mood' and 'anxiety' disorders and not simply postpartum depression. Enough cannot be said about the ability of anyone with a white coat and a medical title to convince vulnerable pregnant women and new mothers that the thoughts and feelings they experience on any given day might be abnormal."

One can clearly see the susceptibility to drug abuse by new mothers whose mental equilibrium is already shaky because of the rigors of childbirth, and especially first-time mothers concerned over the health of both their child and themselves.

Many people feel the drugs and vaccines being administered to children have not been fully tested or guaranteed safe. They feel children are being used as guinea pigs for Big Pharma and that such indiscriminate drugging amounts to nothing less than chemical child abuse.

Health Ranger Mike Adams pondered, "I often wonder when the rest of the country will wake up and notice that the mass-drugging of our nation's children has gone too far. Why isn't the mainstream media giving this front-page coverage? Why aren't lawmakers demanding an end to the chemical abuse of our children? Why isn't the FDA halting these trials on toddlers out of plain decency? You already know the answer: Because they're all making money from this chemical assault on our nation's children. The doctors, hospitals, drug companies, psychiatrists and mainstream media all profit handsomely from the sales of mind-altering drugs to children. Ethics will never get in the way of old-fashioned greed." Such drugging of future generations also advances the globalists' goal of societal control.

Adams and many other concerned parents say children should be given more sunshine, playtime, and access to nature rather than drugs. They say this is the historically proven and commonsense approach to producing balanced, healthy children.

"But psychiatry has no common sense," argued Adams, "and no one in the industry dares mention that most so-called mental disorders are really just caused by nutritional imbalances. Because to admit to the truth about the mental health of children would be to render their careers irrelevant."

The ultimate issue is deciding where to draw the line on which behaviors are normal and which require some sort of intervention. Bipolar disorder, a psychiatric diagnosis describing persons, usually children, who display a wide range of emotions, from exuberant highs to depressing lows, is seen by some as merely the normal ups and downs of the growth process. Critics are quick to note that there is no scientific means to confirm a diagnosis of bipolar disorder.

In a 2009 interview in *Psychology Today,* David Healy, former secretary of the British Association for Psychopharmacology and author of *Mania,* a book on bipolar disorder, acknowledged that this disorder is somewhat of a mythical entity. "The problems that currently are grouped under the heading 'bipolar disorder' are akin to problems that, in the 1960s and 1970s, would have been called 'anxiety' and treated with tranquilizers or, during the 1990s, would have been labeled 'depression' and treated with antidepressants," he said.

Healy described claims that imbalances of enzymes such as serotonin (a gastrointestinal neurotransmitter enzyme believed to promote feelings of well-being) to explain mood swings as unsubstantiated "biobabble." "What's astonishing is how quickly these terms were taken up by popular culture, and how widely, with so many people now routinely referring to their serotonin levels being out of whack when they are feeling wrong or unwell," he said. Some feel it is fortunate there were no synthetic psychiatric drugs available to dull the moods of the master artist Vincent van Gogh, who some historians believe suffered from bipolar disorder.

And even for those who lack van Gogh's genius but exhibit some behavioral issues, the drugs they are being prescribed are leading to disastrous effects. A 2014 study by researchers from the Bloomberg School of Public Health found that prenatal exposure to selective serotonin reuptake inhibitors was associated with autism spectrum disorder (ASD) and developmental delays (DD) in boys. Nearly one thousand mother-child pairs were included in the population-based case-control study, which also found that young boys were more likely to be affected by the development problems than girls. "We found prenatal SSRI exposure was nearly three times as likely in boys with ASD relative to typical development, with the greatest risk when exposure took place during the first trimester," reported Dr. Li-Ching Lee, Ph.D., a psychiatric epidemiologist in Bloomberg's Department of Epidemiology. This research illustrates the challenge for women and their physicians to balance the risks versus the benefits of taking SSRI medications.

With everyone from infants to senior citizens being drugged for both real and imagined mental disorders, it is troubling to note that one group in particular has been targeted for selective drugging—our military service personnel.

CHAPTER 8

DRUGGING THE MILITARY

EACH MEMORIAL DAY, THE UNITED STATES IS INUNDATED WITH patriotic zeal: war documentaries, ceremonies, and speeches honor the men and women of the military services. Politicians and corporations expound on their support for our veterans.

Yet, contrary to all this flag-waving, not to mention President Obama's pledge that he "will not stand" for mistreatment of veterans on his watch, the lives of vets are fraught with inattention and insecurity, even harassment. Large numbers of vets are impoverished, unemployed, struggling with depression, and experiencing substandard treatment by the Department of Veterans Affairs (VA). One third of all homeless Americans are veterans. And these days, we can add one more troubling fact to this list: many military veterans are being prescribed questionable drugs.

In order to understand the prescription drug scandal as it relates to the military, we must first look at the broader health care issues facing many veterans. First of all, many of them lack access to even basic medical services. In mid-2014, an audit of 731 medical facilities by the VA revealed that more than fifty-seven thousand new patients had been waiting more than ninety days for an initial appointment. About sixty-four thousand who had been enrolled in the system for up to a decade still had not seen a doctor.

The long delays veterans face exacerbate many existing problems, and in some cases even lead to death. Marine Gunnery Sergeant Jessie Jane Duff, a member of the organizing committee at Concerned Veterans for America, said that while the government has admitted to about forty deaths in the Phoenix area, with dozens more facilities are under investigation, the real number is much higher.

"Let's go to the backlog that they had. Fifty-three veterans died a day just waiting on their benefits in 2011," said Duff. "The VA itself has those numbers. We're talking about egregious mismanagement, a culture of corruption that was allowing all these executives to give the impression that they had fourteen days of waiting time, not months and months of waiting time, so they could get bonuses. So I expect it will be several hundred, if not thousands."

She gave an example of veterans in Albuquerque suffering from such ailments as gangrene, heart disease, and brain tumors. Yet these veterans were forced to wait more than four months for treatment. Even for basic treatments, taken for granted by many Americans, veterans are forced to jump through hoops to receive care. Duff reported that in Harlingen, Texas, the VA decided that servicemen had to come back for three separate screenings before they qualified for a colonoscopy.

"What disappoints me the most out of this is that it was deliberate. I used to think it was just mismanagement. I've been reporting on mismanagement for the past year. Now I realize it was all deliberate and it was all in the name of an almighty dollar," said Duff. "I'm so shocked and saddened to know that executives at the highest level were training their employees to hide numbers, training their employees to make it look like veterans were only waiting fourteen days."

Duff decried efforts to throw more tax dollars at the VA, stating, "They have a $150 billion budget. They requested $160 billion for the next fiscal year. They've never been denied anything from the Senate or the House, as far as their budget goes." In Phoenix, a mere 39 percent of this budget goes for actual medical costs, she reported, with 52 percent going for administrative costs, including the purchase of expensive office furniture. She added another $6 million was spent on a sparsely attended national conference in Orlando, Florida.

"They've wasted thousands and thousands and millions of dollars. The money is simply being mismanaged," declared Duff.

Further revelations have unmasked the full extent of mismanagement at the VA. In 2014, the scheduling clerk at the Phoenix VA told CNN that for the better part of a year she was ordered by superiors to manage a "secret waiting list" of those vets seeking medical attention, and would remove the "deceased" notice on dead patients to conceal the number of veterans who died while awaiting treatment.

"What makes the VA scandal different is not only that it affected people at their most desperate moment of need—and continues to affect them at subpar facilities," wrote *Slate*'s chief political correspondent John Dickerson. "It's also a failure of one of the most basic transactions government is supposed to perform—keeping a promise to those who were asked to protect our very form of government. The growing scandal points out more than just incompetence. When the wait times were long and those promises were being broken to veterans, administrators then lied about it. It appears this was true across the country."

John Whitehead, author of *A Government of Wolves: The Emerging American Police State,* notes the U.S. government has been breaking its promises to the American people for a long time now, even as service personnel pledge to uphold and defend the Constitution. "Yet if the government won't abide by its commitment to respect our constitutional rights to be free from government surveillance and censorship, if it completely tramples on our right to due process and fair hearings, and routinely denies us protection from roadside strip searches and militarized police, why should anyone expect the government to treat our nation's veterans with respect and dignity?" he asked.

Former presidential candidate Ron Paul saw the scandal in even grander terms. "We should remember that though the VA's alleged abuse and neglect of U.S. veterans is scandalous, the worse abuse comes from a president and a compliant Congress that send the U.S. military to cause harm and be harmed overseas in undeclared, unnecessary, and illegal interventions."

In mid-2014, the uproar over the VA scandal prompted the resignation of Veterans Affairs Secretary Eric Shinseki and an unusual bipartisan proposal from Congress. Vermont independent senator Bernie Sanders and Arizona Republican senator John McCain suggested legislation to correct the VA problems. The bipartisan bill would provide for twenty-six new VA medical facilities in eighteen states and would provide $500 million to hire more VA doctors and nurses. In a two-year trial project under this bill, veterans would be allowed to seek private medical help if they experienced long wait times or lived more than forty miles from a VA facility.

But the legislation did little to counteract many other issues with the program: long waiting times and inferior medical treatment are not the only concerns. As far back as 2009, the FBI launched Operation Vigilant Eagle, a program warning of "militia/sovereign-citizen extremist groups," to include veterans from Iraq and Afghanistan. As a result of the operation, many vets have been targeted for surveillance, censored, threatened with incarceration or involuntary commitment, labeled as extremists and/or mentally ill, and stripped of their Second Amendment rights. There have been instances of veterans having weapons confiscated after answering yes on questionnaires asking if they own firearms. Government policy characterizes veterans as potential domestic terrorists because they might be "disgruntled, disillusioned or suffering from the psychological effects of war."

This demonization of vets has resulted in inappropriate uses of force on the part of many police departments. One example is the case of Jose Guerena, a Marine who served two tours in Iraq. Guerena was killed in 2011 after an Arizona SWAT team kicked open the door of his home during a mistaken drug raid and opened fire. Guerena had no prior criminal record and the police subsequently found nothing illegal in his home.

In 2014, John Edward Chesney, a sixty-two-year-old Vietnam veteran, was also killed by a SWAT team, which responded to a call that the army veteran was standing in his apartment window waving what looked like a semiautomatic rifle. Instead of attempting to make contact with the man, or figure out what he was holding, SWAT officers fired twelve rounds into Chesney's apartment window. It turned out that the gun Chesney was pointing was a "realistic-looking mock assault rifle." In a situation similar to Chesney's, though with a less disastrous conclusion, another Iraq war veteran, twenty-five-year-old Ramon Hooks, was arrested and charged with "criminal mischief" after a Homeland Security agent reported him as an active shooter, even though Hooks was merely practicing with a pellet gun.

The list goes on. In 2012, Brandon Raub, a twenty-six-year-old decorated Marine, was arrested by government agents and held against his will in a psychiatric ward for expressing his views on government corruption on Facebook. The crime for which the vet was isolated from his family, friends, and attorneys seemed to be that he held "conspiratorial" views about the government.

Against the background of this widespread mistreatment of veterans, it's no surprise that when it comes to medication and drugs, they face additional problems. Many of these medications and treatments make veterans' health problems after returning home even worse. The vaccines routinely administered to U.S. troops are one issue of concern, in particular the fact that these vaccines contain squalene. Squalene is an organic compound originally derived from shark liver oil and used as an adjuvant, or additive, to immunization vaccines. An oil molecule native to the body, squalene is found in trace amounts throughout a person's nervous system and brain. What differentiates "good" from "bad" squalene is the route by which it enters your body. Injection is an abnormal entry route, which causes the immune system to attack all the squalene in your body, not just the vaccine adjuvant.

For years, the Department of Defense denied the presence of squalene in the anthrax vaccine. However, the FDA has tested several samples of the vaccine and found the compound present in varying levels. This late admission irritated the late Washington State representative Jack Metcalf, who complained, "We've been told for three years there is no squalene in the anthrax vaccine, then suddenly we are told, 'Oh yes, it's there, but it's no big deal: it's everywhere.'"

Citing the Military Vaccine Resource Directory website, Dr. Anders Bruun Laursen, who has written extensively on vaccines in general and squalene in particular, noted, "The average quantity of squalene injected into the U.S. soldiers abroad and at home in the anthrax vaccine during and after the Gulf War was 34.2 micrograms per billion micrograms of water. According to one study, this was the cause of the Gulf War syndrome in 25 percent of 697,000 U.S. personnel at home and abroad."

Dr. Laursen said squalene "in all probability was responsible for the Gulf War syndrome" and this has engendered "a deeply rooted mistrust in our politicians and the vaccine producers' motives and morals."

One vaccine allotment contained 10.68 milligrams of squalene per 0.5 milliliter, which corresponds to 2.136.0000 micrograms per billion micrograms of water; this is one million times more squalene per dose than advised in the Military Vaccine Resource Directory. "There is [every] reason to believe that this will make people sick to a much higher extent . . . This appears murderous to me," said Dr. Laursen.

Professor Robert F. Garry confirmed these squalene levels in his testimony before the House Subcommittee on National Security, Veterans Affairs and International Relations in 2002. Garry was the first to discover the connection between the Gulf War syndrome and squalene. Unsuccessful efforts have been made to ban squalene from vaccines but controversy continues to rage.

At a 2010 gathering of the American Rally for Personal Rights in Chicago, retired air force captain Richard Rovet, who also is a registered nurse, warned that squalene MF59 was forced on all servicemen beginning in 1999 via the mandatory anthrax vaccine. He said the adjuvant caused many of his comrades to suffer severe and permanent side effects. He said one of his close friends died from it.

"For the past sixty-four years, the United States Military and other agencies within our government have used our servicemen and women as test subjects, oftentimes in secret and without informed consent," explained Captain Rovet.

Rovet noted that in December 1994, the United States Senate released a report titled "Is Military Research Hazardous to a Veteran's Health? Lessons Spanning Half a Century," which outlined the unethical use of servicemen and women as test subjects—human guinea pigs.

Exposing soldiers to harm off the battlefield is nothing new. Some four hundred thousand U.S. soldiers suffered the effects of debilitating amounts of radiation during nuclear bomb testing between 1945 and 1963. And the results of the Public Health Service's infamous Tuskegee syphilis experiments are well documented. In this experiment, which lasted between 1932 and 1972, nearly four hundred black Americans were studied for the effects of syphilis but never treated. Termed an "outrage" and "profoundly morally wrong," President Bill Clinton publicly apologized for the experiments in 1997.

Although the current drugging of soldiers is less well documented, the results may be even more devastating. As reporter Alex Jones put it, "The mass drugging of U.S. troops is one of the most underreported scandals of the modern era, with soldiers not only being used as guinea pigs in a brave new world of pharmacological experimentation, but also having their rights stripped as a result." Michael's House, a Palm Springs drug treatment facility, reported that since 1999, more than seventeen thousand soldiers have been discharged from the U.S. military due to drug use. In that same time span, the number of failed drug tests in the air force has increased 82 percent, and in the army 37 percent.

As the *New York Times* confirms, recent years have seen a significant increase in stimulant use in the military, with annual spending on these drugs rising from $7.5 million in 2001 to $39 million in 2010. According to data provided by Tricare Management Activity, which manages health care services for the Department of Defense, the number of Ritalin and Adderall prescriptions written for active-duty service members increased by nearly 1,000 percent, jumping from three thousand to thirty-two thousand between 2007 and 2012.

More prescriptions have meant more deaths, many of them self-inflicted. Between 2001 and 2009, as the number of psychiatric drug prescriptions among active troops rose 76 percent, the suicide rate increased more than 150 percent. "These soaring statistics cannot be attributed to the horrors of war, as 85 percent of military suicide victims had never even seen combat," noted the Citizens Commission on Human Rights (CCHR). "This suggests that the PTSD [post-traumatic stress disorder] diagnosis is being widely handed out to active-duty and vets to justify putting more and more of them on cocktails of prescribed mind-altering drugs from which they may never recover."

In past years, war trauma, then called "shell shock," was treated with compassion, understanding, and love. But according to the CCHR and many others, today, the willingness to empathize with the warrior and listen to his experiences has been replaced by a psychiatric pop-a-pill quick-fix mentality that employs antidepressants, antipsychotics, stimulants, sedatives, or antianxiety drugs that may produce harmful consequences.

According to a report from Veterans for America, "U.S. troops are being forced to take drugs like Prozac and Seroquel for anxiety and depression. Troops cannot refuse to take the drugs without consequences from their superiors."

According to a Defense Department of Defense Directive in 2011 entitled, "DOD Patient Bill of Rights and Responsibilities in the Military Health System (MHS)," military personnel are entitled to informed consent for any treatment and may refuse treatment. However, many times soldiers are led by suggestion or innuendo to believe that they cannot refuse mental health treatment.

But while there is no permissible enforced treatment for active-duty personnel, veterans may be threatened with losing benefits if they refuse psychiatric treatments recommended by VA hospitals or clinics.

According to some veterans groups, the unprovoked wars in the Middle East (recall that most of the 19 named hijackers of 2001 attacks were Saudi Arabians) have seen instances of resistance by U.S. troops to their superiors' orders. Some military personnel are upset because they must bear the brunt of local vengeance in the wake of atrocities carried out by the growing number of civilian contractors who wear very much the same uniforms as servicemen but are paid more than double their salaries. It has been claimed that sometimes patrols decline to carry out their "search and kill" missions and, instead, return to their bases claiming they carried out their orders.

There's no doubt that post-traumatic stress disorder is a serious problem in the military, but it's also apparent that the new and dangerous drugs being prescribed to soldiers are only making the problem worse. Stimulants help troops stay awake and alert but also contribute to PTSD. Such stimulants generate norepinephrine, an adrenaline-type chemical, which can create vivid and long-lasting traumatic memories. "Because norepinephrine enhances emotional memory, a soldier taking a stimulant medication, which releases norepinephrine in the brain, could be at higher risk of becoming fear-conditioned and getting PTSD in the setting of trauma," wrote Richard A. Friedman, a professor of psychiatry and director of the psychopharmacology clinic at Weill Cornell Medical College.

And many soldiers find it difficult to kick drug habits acquired with the support of the military when they return to civilian life. Often addicted to prescription medications, upon returning home, veterans are forced to visit psychologists who diagnose them with mental disorders and continue to prescribe them drugs, often with tragic consequences.

The case of Andrew White is just one example of the tragic consequences of the overuse of psychotropic drugs among returning U.S. soldiers. White, a twenty-three-year-old veteran of the Iraq War, died in 2008 of an overdose of Seroquel, Klonopin, and Paxil, all prescribed by VA doctors. The death was doubly painful for his parents, Stan and Shirley White, who had already lost their oldest son, Bob, a paratrooper killed earlier in Afghanistan. The Whites blamed Andrew's death on overmedication while under the care of both government and private doctors.

White's mother said she was carefully administering Andrew's daily medication intake. Andrew's father, Stan, remarked that the family now refers to the three drugs Andrew had been taking as the "lethal cocktail." Stan commented, "It's antidepressants, antipsychotics, and analgesics. It's just overloading, and your body can't take it." His wife added, "He was taking exactly what the VA told him to take . . . He made that choice to trust the VA and that trust cost him his life."

A spokesperson told ABC News that the VA investigated Andrew White's death and concluded his doctors had met "the community standards of care," and so had done nothing wrong. Still, the army's surgeon general's office admitted it was working toward better communication between soldiers, their families, commanders, and health care specialists.

Andrew White is not the only soldier who has died from PTSD medications. At least three other war veterans died in their sleep within weeks of each other in the same part of West Virginia in 2008. San Diego neurologist Dr. Fred Baughman became intrigued by the unusual deaths. "Young men in their twenties don't just die in their sleep," he reasoned. Despite no access to medical files, Baughman compiled a list from news reports of some three hundred military deaths linked to sudden heart attacks. The common thread between the hundreds of deaths was the use of antipsychotics and antidepressants, which, depending on type and quantity, are known to cause sudden cardiac arrest.

When not causing physical harm, these drugs have also been connected to severe emotional distress, the very issue against which they allegedly protect. North Carolinians John and Mary Nahas found their son, Iraq War veteran Michael Nahas, bleeding in the bathtub from a suicide attempt. The Nahas believed his behavior was the result of a toxic blend of prescription drugs, including Oxycodone, Xanax, Percocet, Klonopin, Celexa, Lunesta, and Ambien. "We noticed a decline in his personality from the drugs," explained Mary. "They change cognition and behavior. We noticed anger, [he] just couldn't think straight."

In 2013, the Pentagon released a study of military suicides entitled "Risk Factors Associated with Suicide in Current and Former U.S. Military Personnel." This study was based on current and former soldiers' responses to a variety of questions. However, while the study concluded that mental health problems, including manic-depressive disorder, depression, and alcoholism, played a major role in military suicides, none of the questions specifically covered prescribed psychiatric drugs. In fact, the words "drugs" and "medication" do not appear at all in study's questionnaire.

Soon Big Pharma may even offer a drug specifically tailored to PTSD. Researchers at the Massachusetts Institute of Technology, in conjunction with Massachusetts General Hospital, are working on a vaccine to block ghrelin, a stomach hormone produced by the body in response to stress. They found that when injected with drugs to block the excess production of ghrelin, rats appeared to have fewer symptoms associated with PTSD than those not given the drug. But another study indicated that blocking the production and uptake of ghrelin might inhibit the body's ability to regulate energy balance and food intake, which could lead to obesity. Again, we see that these medications can have dangerous side effects. And it's fair to ask whether another drug treatment is really needed. Some claim this new drug, like many others, attempts to correct an underlying health condition by merely covering up its symptoms, rather than attacking the root cause.

In the United States of America's death culture, drugs are a huge part of the problem. In conjunction with the toxicity of many other aspects of American life, they have contributed to numerous deaths and public health issues.

CHANGING THE GAME

AMERICA MUST REVAMP BOTH THE LAWS AND CITIZENS' ATTITUDES regarding drugs before the entire population is zombified by chemical substances. The only way the misuse of drugs can be effectively addressed is by identifying and addressing the underlying causes of drug abuse: poverty, wealth inequality, and hopelessness. Obviously, harsh and discriminatory prosecution at the federal level has proven ineffective. Instead, such an upheaval in thought and action must take place at the local level, for only at the local level can the suitable remedies for this problem be found.

When faced with a medical problem, thoughtful citizens must

not depend upon only one source of information. Seek second opinions from health authorities and study for yourself the evidence available in books and on the Internet. The person who cares most about your health is yourself. Too often, conventional medicine is based on outmoded training, misleading marketing, and the drive for profits. Be wary of government-driven hysteria over potential pandemics as well as government agencies that have proven more beholden to corporate interests than concerned for the well-being of the public. Keen attention must be given to legal prescription drugs as well as the drugs we have deemed illegal. Individuals should educate themselves on the benefits of drugs, but also on the dangers of usage and the risk of abuse. Serious attention should be given to preventive measures such as nutrition, diet, and exercise, rather than simply the possibility of chemical treatment.

The question of vaccines especially requires serious consideration. While some vaccines have been proven effective against disease, the increasing use of adulterants such as thimerasol and squalene makes many current vaccines dangerous. Vaccines should be scheduled over a longer period of time to avoid the medical issues associated with simultaneous vaccines. Public funds should go for the study of vaccines by truly objective organizations, not by the vaccine manufacturers. Laws protecting vaccine makers should be revoked and individual liability returned to those producing the drugs. Government protection agencies such as the Food and Drug Administration (FDA) and Child Protective Services (CPS) must be reorganized to avoid even the appearance of conflict of interests.

It is long past time that marijuana, an herb with no record of physical harm yet proven health benefits, be regulated rather than outlawed. Prohibition has led to the prosecution of hundreds of thousands of otherwise lawful citizens, and has proven no more successful than alcohol prohibition. Thousands of lives have been ruined and public funds wasted on these inequitable marijuana laws, while criminal enterprises have only expanded.

The time has come to recognize psychiatric mind-altering drugs as the driving force behind the horrific rise of public shootings and the increasing suicide rates among teens and veterans. The corporate mass media, so dependent on drug advertising, should shift the focus of its ire from guns to the drugs that clearly cause both suicidal and homicidal tendencies. It would appear that the ongoing agenda of government and corporate media concerning firearm fear mongering has more to do with disarming the population than protecting it.

Like society in general, the military should look past the surface problem of drug use by its soldiers to the underlying causes. It is no wonder that troops sent to faraway locations to fight unnecessary wars for corporate interests should turn to drugs to relieve their pain. After being trained to kill and given indiscriminate amounts of mood-altering drugs while in service, they are then largely abandoned and even branded as potential terrorists by their own government when they return home. Once our troops are through defending true freedom and liberty, they should find the dignity and respect they so deserve.

CHAPTER 9

DEADLY FOOD

As the old saying goes, "You are what you eat." If this is true, it bodes ill for many Americans.

Globalist-controlled corporations have turned our food poisonous. It is laced with sodium nitrate for color, the nonessential amino acid monosodium glutamate (MSG) for taste, and various chemicals to preserve shelf life. Processed food has been intentionally stripped of essential minerals, vitamins, and nutrients in a globalist plot to promote disease and malnutrition. Some nutritionists believe that this corporate-sponsored genocide from deadly food will ultimately lead to large-scale social collapse.

Many self-styled globalists make no secret that they believe the world is suffering from overpopulation and that a steep reduction of the planet's inhabitants is necessary to ensure a future without hunger and civil unrest.

"The food supply appears to be intentionally designed to end human life rather than nourish it," claimed nutrition advocate Mike Adams. "After having now analyzed over one thousand foods, super-foods, vitamins, junk foods and popular beverages for heavy metals and other substances at the Natural News Forensic Food Labs, I have arrived at a conclusion so alarming and urgent that it can only be stated bluntly. Based on what I am seeing via atomic spectroscopy analysis of all the dietary substances people are consuming on a daily basis, I must now announce that the battle for humanity is nearly lost."

He stated that this adulteration of our food is not happenstance or simply a mistake. "My lab has uncovered scientific proof that substances are intentionally formulated into dietary products to *drive consumers mentally insane* [emphasis in the original] while causing widespread infertility, organ damage, and a loss of any ability to engage in rational, conscious thinking," wrote Adams. "These toxic substances are being found across the entire food supply including in conventional foods, organic foods, 'natural' products, and dietary supplements."

Adams said such intentional formulations go far beyond the contamination of foods with heavy metals and include the intentional inclusion of toxic substances in products for mass consumption. "The result is what you see unfolding around you right now: mass insanity, incredible escalations of criminality among political operatives, clinical insanity among an increasing number of mainstream media writers and reporters, widespread infertility in young couples, skyrocketing rates of kidney failure and dialysis patients, plus a near total loss of rational thinking among the voting masses," he said.

Adams and others argue that widespread food poisoning may cause the collapse of a capable workforce, the rise of masses dependent on government for survival, the collapse of free democracies due to the cognitive deficiencies of the voting masses, an exploding prison population and the rise of for-profit corporate prison systems, and even the near complete collapse of any ability of the news-consuming public to parse and comprehend the most basic information such as national debt figures.

An estimated one in three Americans today is obese. In addition to sugary processed foods lacking in nutrients, researchers in recent years have found that a notable cause of this national weight gain is due to declining physical activity in the workplace.

While diet, lifestyle, and genetics all play a role in the rise of obesity, a 2011 study published in the journal of the *Public Library of Science* (*PLOS*), reported that jobs requiring moderate physical activity accounted for 50 percent of the labor market in 1960. By 2012, this number had dropped to just 20 percent. The remaining 80 percent of jobs are sedentary or require only light activity, accounting for an average decline of 120 to 140 calories a day in physical activity, which closely matched the population weight gain in recent years. These findings pose a challenge to employers to heighten workplace health initiatives and pay more attention to physical activity at work.

According to a report by the New England Complex Systems Institute (NECSI), the ongoing conversion of corn crops to ethanol has contributed to a doubling in global food prices since 2005. Corn forms the basis for everything from high-fructose corn syrup and cereal to feed for livestock. Dr. Yaneer Bar-Yam, founding president of the NECSI, said there was a link between global food-price increases associated with the corn-to-ethanol conversion and the violence in the Middle East and North Africa known as the Arab Spring.

He said the amount of corn used to produce one gallon of ethanol fuel could feed one person for a day, and the U.S. diversion of corn for ethanol could feed as many as 570 million people worldwide annually. "When you're pulling into the gas station and you're filling your tank with gas, 10 percent of what you're putting into the tank is food. It could be eaten by people instead," said Bar-Yam.

Fed Up, a 2014 documentary, revealed more sobering facts about the U.S. food industry. Within two decades, the documentary predicts, more than 95 percent of all Americans will be overweight or obese, and one out of every three Americans will have diabetes by 2050. The film pointed out that 80 percent of all food items sold in America have added sugar. This greatly contributes to the epidemic of obesity.

One would think that such terrifying projections would spur some sort of legislative reform. Yet a corporation-controlled Congress has been unwilling to make the necessary changes. In November 2011, House Republicans, spurred on by the food corporations that supply the nation's school cafeterias, produced a spending bill barring the USDA from changing its nutritional guidelines for school lunches. The proposed changes would have required more green vegetables and less sodium and sugar. Instead, the bill that passed has such absurdities as a provision classifying tomato paste on pizzas as a vegetable, although tomatoes are actually a fruit.

But the health concerns go beyond pizza. Giant food corporations sell processed food designed to keep consumers addicted to their products, not to keep them nourished. The food industry is no different from any other commercial enterprise—their products and marketing are geared toward maximizing profits rather than benefiting humankind. Food products chemically processed with refined ingredients or artificial substances are deliberately designed to be deficient in nutrients, which leaves consumers craving more and more of the products.

And the evidence is clear that these processed foods are responsible for deadly conditions such as diabetes and heart disease. One of the main contributing factors to such illnesses is sugar, which is as addictive and harmful as many illicit drugs. In her book *Suicide by Sugar: A Startling Look at Our #1 National Addiction,* author Nancy Appleton pointed out that sugar can suppress the immune system, elevate glucose, heighten insulin responses, cause hyperactivity and anxiety, and raise triglyceride levels. It has been noted that the obesity epidemic in the U.S. generally began about 1977, the same year that the first dietary guidelines for Americans were published. These dietary guidelines included restrictions on fat, while largely giving carbohydrates and sugars a free pass. The tremendous health risks posed by excessive carbohydrate consumption are now clear.

In America, even natural fruit juices contain almost as much sugar as soft drinks, and the increased use of sugar has resulted in debilitating effects on both mind and body.

Processed foods tend to be unhealthy because they include high levels of sugar, or even worse, high-fructose corn syrup (HFCS), a synthesized sweetener commonly found in breads, cereals, breakfast bars, lunch meats, yogurts, soft drinks, soups, and condiments. HFCS has been linked to the world's leading killers—heart disease, diabetes, obesity, and cancer—and activates the same areas in the brain as highly addictive drugs like cocaine. A 2007 Rutgers University study compared HFCS sodas with those sweetened with traditional sugar (sucrose) and found the HFCS drinks contained up to ten times more harmful carbonyl compounds—substances previously linked to serious health complications in diabetics.

In an effort to increase sales, food manufacturers engineer processed foods that are sweet, salty, and fatty, all flavors the body naturally craves. But while the taste is there, the nutrition and fiber are not, resulting in what has been called the "food reward hypothesis of obesity."

It is the globalist bankers and owners of the large multinational food corporations who profit from a sugar-addicted and sedated population. They are achieving their goal of population reduction while profiting from the increasing number of people in their health care industry.

As Kris Gunnars explains, "Food manufacturers spend massive amounts of resources on making their foods as 'rewarding' as possible to the brain, which leads to overconsumption . . . Some people can literally become *addicted* [emphasis in the original] to this stuff and completely lose control over their consumption." Though rarely discussed by the mass media, food addiction, in which brain biochemistry is altered by eating processed foods, is the primary reason why so many people are unable to stop unhealthy eating habits.

Gunnars also points out that while controversy rages over how many carbohydrates we should consume, too few nutrition pundits are distinguishing between the different types of carbohydrates. "The carbohydrates you find in processed foods are usually refined, 'simple' carbohydrates. These lead to rapid spikes in blood sugar and insulin levels and cause negative health effects." He even warned against some "whole grain" products, stating, "These are usually whole grains that have been pulverized into very fine flour and are just as harmful as their refined counterparts."

Consumption of refined carbohydrates can have disastrous effects on our general health. A 2010 *Food & Nutrition Research* study showed participants who ate a processed food sandwich (white bread and artificial cheese) absorbed only half as many calories as those who ate a whole food sandwich (multigrain bread and cheddar cheese).

Many processed foods also contain refined seed and vegetable oils, which are often hydrogenated. According to numerous scientific studies, people who consume large amounts of these oils have a significantly increased risk of heart disease, the most common cause of death in Western countries. Instead, healthy consumers should be replacing processed oils and trans fats with natural fats such as olive oil.

When we're not creating our own harmful foods products, we're adulterating existing natural foods. Beef is a staple of the American diet. Despite the fact that many Americans eat beef several times a week, most have no idea that the feedlot cows they're consuming subsist on a diet of garbage and waste products. Cattle are often fed "poultry litter," the agriculture industry's term for the waste picked off the floors of chicken cages. Consisting of feces, feathers, and uneaten chicken feed, a typical sample of poultry litter may also contain antibiotics, heavy metals, disease-causing bacteria, and even rodent carcasses, according to Consumers Union, the nonprofit organization that publishes *Consumer Reports*.

It is not just cows and chickens that are being fed poop. A greater and greater percentage of the seafood eaten by Americans—currently 35 percent—is sourced from Asia, and these shipments are frequently contaminated. The FDA, ostensibly the barrier between contaminated food products and the U.S. market, admits that it inspects only about 2.7 percent of imported food. Since 2007, FDA inspectors, despite the tiny percentage they see, have rejected 1,380 loads of seafood from Vietnam for filth and salmonella and 820 shipments from China.

In 2012, the FDA in effect barred Moon Fishery (India) Pvt. Ltd. from importing fish to the USA because of contamination and unsanitary conditions at its plant. But before instituting the ban, a salmonella outbreak in twenty-eight states had sickened hundreds of Americans. Considering such numbers and events such as the tardy fishery ban, it is apparent that the FDA either cannot, or will not, provide adequate oversight protection of the public's food.

With so many food products tarnished by toxic compounds or grown in disgusting conditions, more and more grocery shoppers today are turning to organic foods. Organic food and fiber sales in 2013 grew to $35 billion, an increase of 11.5 percent from 2012. Even large chains, such as Walmart and Target, have increased their offering of organic products.

"Organic is booming, and the mainstream acceptance of organic products is driving it," said Steve Crider, a spokesman for Amy's Kitchen, a California-based organic and natural food maker. "We are coming into the second generation of organic consumers: the kids who were raised on this stuff by their moms. They get it about food and sustainability and organic and local. They are part of the drivers of this."

However, there are many problems even with organic food. This is partly because it has become increasingly easy for a food to be certified organic. Much as some "whole grain" products have become just as unhealthy as their refined counterparts, the term "organic" just doesn't mean much anymore. Regulators continually water down the requirements to gain the much-coveted seal of approval from the USDA, making it hard for consumers to discern which products are actually grown under organic conditions.

The National Organic Standards Board (NOSB), which determines which products are certified organic, originally operated under "sunset" provisions stipulating that organic products using nonorganic materials be dropped after five years unless renewed by a two-thirds vote of the board. But in 2013, a rule change reversed this process, keeping certification for these products in place unless a two-thirds vote removes it. Yet given the makeup of the NOSB, which is stocked with food industry reps seeking to sell more of their dubiously "organic" products, any vote to rescind organic status is exceedingly unlikely. This policy change means that consumers will find it more difficult than ever to determine which products are truly free of harmful ingredients.

Quality Assurance International (QAI) is North America's largest for-profit organic certifier and typifies the work of nearly one hundred such certifying firms whose job it is to inspect producers, processors, handlers, and retailers seeking the "certified organic" stamp in order to assure they follow organic standards. These standards include verification that irradiation, sewage, synthetic fertilizers, prohibited pesticides, and genetically modified organisms are not used; that antibiotics or growth hormones are not used on animals; that products have 95 percent or more certified organic content.

Operating worldwide, QAI literature states "QAI's programs verify organic integrity at each link of the production chain." But critics say QAI and others may be loosely interpreting these standards, allowing some companies to bend, if not break, the rules.

For example, according to Mark Kastel, cofounder of the Wisconsin-based organic watchdog group the Cornucopia Institute, many large dairy operations, dubbed "factory farms," bend the rules by confining thousands of cows in feedlotlike conditions. Cornucopia charges that QAI, which certifies milk from Dean Foods' Horizon brand and Woodstock Farms' Aurora Dairy, approves their products by broadly interpreting a section of the organic guidelines stating that cows must have "access to pasture" as including feedlot operations. Some milk certified organic by QAI, Cornucopia says, is also fortified with omega-3, a polyunsaturated fatty acid not approved under National Organic Program standards. In February 2007, the Cornucopia Institute sued the USDA, accusing the agency for failure to enforce the law.

Lax organic requirements are largely a legacy of backroom political deals. As Kastel explains, "The industry-friendly regulators under the Bush administration had laws on the books to control these problems, but didn't have the political will to control them," he explained. As long as there's money to be made on organic products, there will always be the risk that regulators will stretch to classify almost anything as organic.

Legitimately harmful materials are found in many foods labeled organic. For example, organic apples at the supermarket may contain streptomycin and oxytetracycline, antibiotics that disqualify meat products from being labeled organic but are still used on apples and pears due to the regulatory quagmire.

And consumers are remarkably ill-informed about what goes into their allegedly organic products. A *Consumer Reports* National Research Center online survey of about one thousand people revealed that 85 percent of respondents either didn't think (68 percent) or didn't know (17 percent) that antibiotics are used to treat disease in apple and pear trees. More than half said they did not think treated fruit should be labeled as organic. The Environmental Protection Agency (EPA) claims the risk of dangerous amounts of antibiotics in organically labeled fruit is quite small. But do we really want to take any chance, especially when government regulators have proven so often that they don't have the public's best interest at heart? Many groceries and markets now are offering organic or free-range beef and poultry. But how does one know the purity of this meat? Many Americans have tried to turn to healthy foods to fight obesity and its attendant health problems. But even then, they are at the mercy of giant food corporations, which have exposed them to all sorts of harmful products.

For example, PowerBar, the first energy bar used by endurance athletes, was developed by a Canadian athlete and a nutritionist in 1986, but later sold to Nestlé, said to be the world's leading nutrition, health, and wellness company, which is headquartered in Switzerland.

A year after Kraft Foods, a North American unit of Altria Group, Inc. (formerly Philip Morris companies), acquired Boca Burger Inc. in 2000, a privately held maker of soy-based meat alternatives, its Florida plant was closed.

In 2001, laboratory analyses by Greenpeace showed Kellogg's-owned Morningstar Farms meat-free corn dogs tested positive for StarLink GMO corn and genetically altered soy not approved for human consumption. After Greenpeace urged the Food and Drug Administration to issue an immediate recall and health investigation of the Morningstar product, Kellogg's voluntarily recalled its meat-free dogs as a "precautionary measure."

Cascadian Farms, an organic food firm begun in 1972 even before the FDA started certifying organic food, is now owned by General Mills.

Health drinks are not immune to corporate control. Naked Juice of California operates as a subsidiary of PepsiCo. Odwalla Inc., a leading maker of fruit juice, smoothies, and food bars, is a wholly owned subsidiary of Coca-Cola.

Nestlé, Kraft, Kellogg's, General Mills, PepsiCo, and Coca-Cola collectively donated nearly $10 million in addition to the $12 million given by Monsanto and DuPont in 2012 to defeat California's Proposition 37 ballot initiative proposing mandatory GMO labeling.

CHAPTER 10

DEADLY GRAINS

THE BIBLE STATES THAT EACH PERSON SHOULD HAVE THEIR "DAILY bread," yet even bread is no longer the healthy, nutritious food many believe it to be.

Many factors contribute to bread products that are increasingly unhealthy and non-nutritious. One culprit is the intensive agribusiness farming practices that produce the wheat used in the bread we eat. Today's wheat has 28 percent fewer health-providing minerals like magnesium, iron, zinc, and copper as compared to wheat of previous generations.

Furthermore, today's bread contains many unsafe ingredients. The *Ecologist,* a United Kingdom news site, featured a study showing widespread contamination of bread by glyphosate, an endocrine disruptor. Glysophate, the basic ingredient of the herbicide Roundup, destroys microorganisms in the human gut and hinders cellular detoxification. Studies have also linked it to cancer, Parkinson's disease, and Alzheimer's.

For some time, travelers have reported no health problems when eating breads and pasta in Europe yet suffering from a variety of illnesses after eating grain food back in the USA. Recent scientific papers have stated that the problem lies in the way American farmers harvest wheat.

Just before harvest, Roundup and other herbicides containing glyphosate are routinely applied to wheat and barley both as a drying agent and to increase the amount of seeds in the stressed plant.

According to a USDA report issued in December 2014, more than half of food (including processed fruits and vegetable and even baby formula) tested by the government agency showed detectable levels of pesticides, though it was stated most were within levels the federal government considers to be safe. USDA statistics, as of 2012, showed 99 percent of durum wheat, 97 percent of spring wheat, and 61 percent of winter wheat as part of the harvesting process had been doused with Roundup, the world's most widely used herbicide.

Probiotics, or beneficial gut bacteria, plays a critical role in human health. Gut bacteria aid digestion, discourage the development of autoimmune diseases, synthesize vitamins, and help build the body's immunity. Many believe that Roundup because of its glyphosate significantly disrupts the functioning of beneficial bacteria in the gut, leading to illness.

In 2013, the EPA approved increased tolerance levels for glyphosate in oilseed crops and vegetables. A 2014 Reuters news article noted that the USDA does not test for residues of glyphosate, the active ingredient in Roundup, A USDA spokesman who asked not to be quoted said that the test measures required for glyphosate are "extremely expensive . . . to do on a regular basis." Unusually high levels of glyphosate have even been found in the breast milk of American mothers, at anywhere from 760 to 1,600 times the allowable limits in European drinking water, while urine tests show Americans have ten times the glyphosate accumulation as Europeans.

Dr. Stephanie Seneff, a senior research scientist at MIT who has produced more than 170 peer-reviewed articles concerning biology and technology, has connected autism to glyphosate toxicity. She said there is a "remarkably consistent correlation" between the use of the herbicide Roundup on soy, corn, and wheat crops and the rising rates of autism.

"At today's rate, by 2025, one in two children will be autistic," she warned a conference audience in late 2014. "In my view, the situation is almost beyond repair. We need to do something drastic."

Dr. Seneff also said additional chemicals in Roundup are untested because they're classified as "inert," yet according to a 2014 study in *BioMed Research International,* these chemicals are capable of amplifying the toxic effects of Roundup hundreds of times over.

The study reported that Roundup herbicide is 125 times more toxic than its active ingredient glyphosate alone because of "formulations"—chemical additives in the herbicide.

"Despite its relatively benign reputation, Roundup was among the most toxic herbicides and insecticides tested. Most importantly, eight formulations out of nine were up to one thousand times more toxic than their active principles. Our results challenge the relevance of the acceptable daily intake for pesticides because this norm is calculated from the toxicity of the active principle alone. Chronic tests on pesticides may not reflect relevant environmental exposures if only one ingredient of these mixtures is tested alone," stated the study.

Sayer Ji, founder of GreenMedInfo.com, said such studies give a good indication of how the agrochemical industry "sets up the public for mass poisoning" by concealing the true toxicity of their chemical formulations and by focusing solely on the risks associated with the active ingredient in their formulations.

Recently, it was reported that thirty-four of forty bread products sampled from Warburtons, a major British bread manufacturer, contained traces of glyphosate. Critics argue that while glyphosate residue amounts were below the EU maximum allowable, no glyphosate should be tolerated at any level.

The use of bromine in bread is also raising alarm. Bromine is used as an anticaking agent or dough conditioner when bread is milled. After fears in the 1980s that too much iodine was present in bread, iodine was replaced with bromine in the bread-making process. Prior to this, bromine was used as a sedative to treat four different types of psychosis.

Bromine has many industrial applications, and is used in flame retardants, pesticides, sanitary products, fumigants, medicines, dyes, and photographic solutions. Ingesting high quantities of bromine, a central-nervous-system depressant, can cause symptoms such as depression, skin problems, and hyperthyroidism. Elevated bromide levels have been implicated in many different thyroid disorders, from simple hypothyroidism to autoimmune diseases to thyroid cancer.

Bromide has also been linked to a variety of neurological problems. Studies have shown that males ingesting four milligrams of sodium bromide daily have decreased attention spans and more frequent feelings of fatigue. This numbing effect could be due to bromide causing oxidation or the loss of electrons in the atoms making up the central nervous system. "In cats, this organic bromine induced REM sleep. Therefore, bromine has a zombifying potential. Why iodine was replaced with a goitrogen possessing carcinogenic and zombifying potentials in a population already very iodine deficient, even by the very low RDA Standard, remains a mystery," wrote Dr. Gary E. Abraham in an article entitled "The Safe and Effective Implementation of Orthoiodosupplementation in Medical Practice" published in the March 2004 issue of the *Original Internist*. "Nevertheless, it is a very effective way to keep a nation sick and zombified," he pointed out.

Bleached flour is a further concern. Many types of bread, especially enriched white bread, should be avoided, according to Dr. Lawrence Wilson, an Arizona physician specializing in nutritional health, because white flour is usually bleached with highly toxic chlorine bleach, the same bleach used to wash clothing. "It is then baked and the concoction creates extremely toxic chemicals in the bread that irritate the intestines," said Wilson, adding that in addition to bromide, iron is added. Abnormally high iron levels have been linked to a variety of diseases, from diabetes to arthritis to cancer.

But don't expect the corporate mass media to warn you of such dangers in your food. Advertisers wield great clout and the government backs them up. Monsanto's growth hormone IGF-1 has been linked to an increased risk of human colorectal and breast cancer in studies both in the U.S. and Canada. However, the Food and Drug Administration (FDA), American Cancer Society, World Health Organization, and National Institutes of Health all have downplayed the significance of such studies.

Award-winning journalists Steve Wilson and Jane Akre were

both fired when they tried to expose the cover-up of such studies as well as reporting on the ban on growth hormones in Europe. According to the Goldman Environmental Prize website, "As investigative reporters for the Fox Television affiliate in Tampa, Florida, [Wilson and Akre] discovered that while the hormone had been banned in Canada, Europe and most other countries, millions of Americans were unknowingly drinking milk from rBGH-treated cows. The duo documented how the hormone, which can harm cows, was approved by the government as a veterinary drug without adequately testing its effects on children and adults who drink rBGH milk. They also uncovered studies linking its effects to cancer in humans. Just before broadcast, the station cancelled the widely promoted reports after Monsanto, the hormone manufacturer, threatened Fox News with 'dire consequences' if the stories aired. Under pressure from Fox lawyers, the husband-and-wife team rewrote the story more than eighty times. After threats of dismissal and offers of six-figure sums to drop their ethical objections and keep quiet, they were fired."

Although mainstream media outlets appear hesitant to produce stories on harmful ingredients in milk and bread, some grassroots efforts have enjoyed greater success in this matter. One example is the controversy over azodicarbonamide (ADA), used as a dough conditioner by large commercial bakers and found in hundreds of sandwich breads and snacks. ADA is also in yoga mats, flip-flops, and is used as a foaming agent in the plastics industry. The World Health Organization has linked ADA with respiratory illnesses and allergies, and the UK Health and Safety Executive (HSE) recognizes ADA as a cause of asthma. The United States is the only country that allows ADA as a bread ingredient, prompting a petition from one concerned citizen, Vani Hari of the nutrition website Foodbabe .com. Hari's petition called on Subway to remove ADA from their bread, pointing out that other industry leaders would likely follow.

The petition worked. Subway executives in 2014 announced that the chain had already begun removing ADA from its bread recipe. Inspired by this success, the Environmental Working Group (EWG), an environmental health research and advocacy organization, pressed other bread makers to do the same. Already, Wonder Bread, and Martin's Famous Pastry Shoppes have pledged to do the same.

Such efforts demonstrate the effectiveness of individual citizen campaigns, about the only thing that *is* effective these days. Protests against other dangerous bread ingredients, such as bromine, will need to be led by similar coalitions of citizens.

Rice is another cereal grain that contains potentially deadly ingredients. In 2012, *Consumer Reports* tested 223 samples of rice products and found high levels of arsenic in most of them, including many products that contained the very toxic inorganic arsenic. The study found that arsenic levels of two hundred parts per billion (ppb) or higher were common even in rice-based baby cereals. Arsenic, historically used as a poison, can be found in both organic and inorganic (metal and mineral) forms.

Naturally, these disturbing findings put pressure on the FDA, which tried to assuage fears that rice products were contaminated with arsenic. The FDA commented, "Agency scientists determined that the amount of detectable arsenic is too low in the rice and rice product samples to cause any immediate or short-term adverse health effects." In failing to note the long-term health risks from chronic exposure to arsenic, the FDA revealed its true agenda—as a policy arm of food corporations rather than protector of the public.

Indeed, the agency has been silent even as compelling evidence about the dangers of arsenic contamination in rice continues to mount. Dartmouth Medical School researchers found that pregnant women eating rice had higher levels of arsenic in their urine than women who didn't. Their study found that eating just half a cup of rice a day could expose a person to as much arsenic as they would ingest if they had been drinking water that contained the maximum contaminant level allowed by law—four parts per million for each liter.

In addition to arsenic, rice protein products in the natural food industry also frequently contain heavy metals such as lead, cadmium, mercury, and tungsten. These metals have even been found in some certified organic foods. By mid-2014, concern over the amount of heavy metals in rice protein products prompted petitions from consumers seeking new limits on these substances.

Like Vani Hari's efforts to limit ADA in bread, the petition drive was an extraordinary success, and resulted in the first voluntary heavy-metals limit ever announced in the natural products industry. Despite the fact that mainstream media coverage of the problem was scant, many major manufacturers responded quickly and positively to the petition, agreeing to reduce heavy metals by July 2015. The proposed new limits stipulated that lead be reduced to .25 ppm (parts per million), tungsten to .05 ppm, cadmium to 1 ppm, and mercury to .05 ppm.

Mike Adams, while praising the metals-reduction efforts of the manufacturers, nevertheless cautioned, "Lead, cadmium and tungsten are still present in every rice protein product we've tested so far, at concentrations that consistently exceed those found in pea protein, hemp protein, whey protein or any other protein source we've tested. While I expect this situation to steadily improve, there is without question still lead, cadmium and tungsten in rice protein products sold on store shelves right now . . . To my knowledge, none of these retailers test the products they carry for heavy metals."

It's difficult for wholesalers and retailers to adequately gauge the safety of the rice products they sell since so many of them come from foreign sources. Nearly all the rice protein used in superfoods and supplements sold in North America still comes from China and other Asian nations where product testing is virtually nonexistent. "To my knowledge, no company offering these products has yet been able to provide its customers an accurate 'country of origin' statement for its rice protein materials," noted Adams. Without basic information like where and how these materials were produced, it's no wonder that so many rice products are marked by contamination.

On a few occasions, the rice issue has emerged into the public consciousness. In early 2014, the FDA issued a recall notice for all shipments of Uncle Ben's Infused Rice in both five- and twenty-five-pound bags shipped in 2013. The recall came after the FDA was alerted to a cluster of illnesses at three public schools in Katy, Texas.

Thirty-four students and four teachers experienced burning, itching, rashes, headaches, and nausea. Uncle Ben's Infused Rice Mexican Flavor made by Mars Foodservices of Greenville, Mississippi, was the common food item all of the afflicted had eaten. The recall echoed a similar case from December 2013. In that instance, the Illinois Department of Public Health notified the CDC of twenty-five children with similar skin reactions following a school lunch that served an Uncle Ben's Infused Rice product. North Dakota also reported a similar incident that occurred in October 2013.

As the Uncle Ben's incidents illustrate, serious contamination scares are repeated because the FDA has failed to take the necessary action to regulate these products. Is the FDA involved in a toxic food cover-up? The very idea used to be called a conspiracy theory. Today, the suspicion is so widespread that even the mass media cover it, though mainstream outlets have failed to explore the possibility that lead and cadmium—even more toxic than arsenic—may also be present in rice.

Could the reluctance of the government, and hence the corporate media, to discuss cadmium in rice be due to the enormous power of large corporate advertisers? Or perhaps because we import so much rice from China, America's top investor? These connections should not be taken lightly.

There is a crisis of contamination in the grains we consume. Yet due to the entrenched interests of multinational food corporations, sensible regulatory policies are unlikely to be enacted anytime soon. As a result, the best way we can effect change is on the personal level: signing petitions, contacting lawmakers, and choosing to eat only grains that we can actually determine are contaminant-free. Yet even when we attempt to make the healthiest choices we can as consumers, we are not always successful. Consider the example of consumers attempting to replace their sugar intake with dangerous artificial sweeteners.

CHAPTER 11

DEADLY SWEETENERS

HEALTH-CONSCIOUS CONSUMERS ARE ALWAYS ON THE LOOKOUT for substances that will sweeten their foods without packing on the pounds. But a close examination of some of the most popular sweeteners reveals deadly ingredients.

TRUVIA

THE SWEETENER TRUVIA CONTAINS AS A PRIMARY COMPONENT erythritol, a sugar alcohol derived from genetically modified corn. Cargill, a multinational food corporation, manufactures Truvia.

It turns out that erythritol is also effective in a much different role: as a pesticide. In mid-2014, an inquisitive grade school student made this disturbing discovery. Simon D. Kaschock-Marenda, the son of an associate professor of biology at Drexel University in Philadelphia, decided to test Truvia on fruit flies as a science project after he learned his parents were trying out various artificial sweeteners in an effort to diet. After purchasing various sweeteners, including Equal, PureVia, and Sweet'N Low, the father-son team fed the flies at Daniel Marenda's Drexel laboratory. To their consternation, they found that after only six days all the flies fed Truvia were dead.

Most of the fruit flies in the study lived between 38.6 and 50.6 days, but the flies fed Truvia fruit had an average life-span of less than one week. "The more you get them [fruit flies] to consume erythritol, the faster they die," noted Drexel biology professor Sean O'Donnell, who along with the father and son and others published a paper on the Truvia study in the journal *PLOS ONE*.

The study found that while the FDA had approved erythritol in 2001 as a food additive, "Our results demonstrate, for the first time, that erythritol may be used as a novel, environmentally sustainable and human safe approach for insect pest control." Yet it's also being used as a food product.

In late 2013, Cargill settled a lawsuit charging that the corporation had misled consumers by marketing Truvia as "natural" even though it contained "highly processed" ingredients and/or GMOs. While denying liability for false advertising, Cargill nevertheless agreed to settle, and announced it would put $5 million toward refunding consumers. It also agreed to change the wording on its product labels, though it would continue to market Truvia as "natural," even though critics said Truvia is actually made from a fermentation process in which yeast organisms are fed GM corn maltodextrin. Cargill has insisted that erythritol is not derived from corn or dextrose feedstock but from yeast organisms. "Yeah, okay, but the yeasts are fed GMOs. So they're playing mind games with their explanations," retorted health advocate Mike Adams. Further inquiry into Truvia is needed, and the early results are certainly concerning. But given the vested interests of large corporations such as Cargill, such rigorous research is unlikely.

SPLENDA

THE POPULAR SWEETENER SPLENDA CONTAINS THE CHEMICAL SUcralose. Sucralose, which is made using chlorine, is six hundred times sweeter than sugar. According to Shane Ellison, known as the People's Chemist, chlorine has a split personality. "It can be harmless or it can be life threatening," he writes. "In combo with sodium, chlorine forms a harmless 'ionic bond' to yield table salt. Sucralose makers often highlight this worthless fact to defend its safety. Apparently, they missed the second day of Chemistry 101—the day they teach 'covalent' bonds. When used with carbon, the chlorine atom in sucralose forms a 'covalent' bond. The end result is the historically deadly 'organochlorine' or simply: a Really-Nasty Form of Chlorine (RNFOC)." Exposure to organochlorines, Ellison argues, can lead to various forms of cancer as well as diabetes.

Sucralose was the subject of a recent study by researchers at the Ramazzini Institute in Bologna, Italy. Institute director Dr. Morando Soffritti and his team fed 843 laboratory mice varying doses of the sweetener throughout the course of their life-spans. Postmortems showed an association between leukemia and sucralose consumption: the more sucralose the mice consumed, the higher their risk of leukemia. Further research is needed, but Soffritti urged pregnant women and children to avoid artificial sweeteners until more extensive tests can be done.

In the wake of the Italian study, the Center for Science in the Public Interest (CSPI) downgraded Splenda in its Chemical Cuisine guide to food additives from "safe" to "caution" pending further review.

Yet the only groups that are conducting further research on Splenda have a spotty record at best: the FDA, Tate & Lyle (the manufacturers of sucralose), and McNeil Nutritionals (the makers of Splenda). Tate & Lyle and McNeil Nutritionals, of course, are incentivized to make sure that their product can be sold. And recall, it was the FDA that approved Vioxx, which led to more than sixty thousand deaths. And Vioxx was a medicine. Splenda is classified as a food additive, which means far less scrutiny from the FDA.

Splenda's manufacturers have defended the drug against the criticisms. Maureen Conway, R.D., director of nutritional affairs for McNeil Nutritionals, points to a recent study led by Dr. Tongzhi Wu at Australia's University of Adelaide School of Medicine and funded by the National Health and Medical Research Council there. Ten men were tested for insulin, sugar, and the hormone GLP-1 after drinking four different drinks following an overnight fast. Conway said the test showed that consumption of sucralose was no worse than water. She said the test "in Australia provides more evidence that Splenda can be used safely by everyone, including pregnant women, children and people with diabetes."

But detractors point out the limited scope of the study. Yes, Splenda is effective in avoiding insulin swings, but this does not prove anything about its long-term health effects in other areas. One potential area of concern is that the aspartame and glutamate used in some artificial sweeteners can act as neurotransmitters in the brain by facilitating the transmission of information from neuron to neuron. Excess aspartame and glutamate may be able to gradually destroy neuronal pathways, causing memory loss, brain lesions, and dementia often well before any chronic illness is apparent.

Furthermore, these sugar substitutes may be less effective than previously thought in keeping weight under control. A 2010 study published in the *Yale Journal of Biology and Medicine* found that noncaloric sugar substitutes, including aspartame and sucralose, do not satisfy the brain in the same way that real sugar does, and consumption of sugar substitutes can lead to overeating and obesity. The paper referred to several large-scale experiments that found links between the use of artificial sweeteners and weight gain. The study also noted that artificial sweeteners can actually encourage sugar cravings and sugar dependence.

Dr. Louis J. Elsas II, director of medical genetics at Emory University School of Medicine, told the congressional Labor and Human Resources Committee that even a single use of aspartame has been shown to raise the level of the essential amino acid phenylalanine in the blood. A neurotoxin, phenylalanine excites neurons in the brain to the point of cellular death, causing emotional and behavior problems. He said excessive amounts of phenylalanine, which constitutes 50 percent aspartame, can damage parts of the brain and is especially dangerous for infants and fetuses.

We're far from knowing everything there is to know about Splenda, but early indications are that it may have unintended consequences and be relatively ineffective as a solution to the problem—an obesity epidemic—it is intended to address.

ASPARTAME

SPLENDA, HOWEVER, COMES OUT LOOKING GREAT IN COMPARISON TO aspartame, described by Dr. Joseph Mercola as "by far the most dangerous substance added to most foods today." Sucralose, which became popular in part as a result of health concerns related to aspartame, has gained market share, but aspartame is still used in many foods and drinks.

Questions abound regarding aspartame, which is now found not only in diet drinks but also in thousands of food products as well as over-the-counter drugs. The sweetener's history is troubling. The Pentagon once listed it as a biochemical warfare agent; then the pharmaceutical company G. D. Searle discovered that aspartame also produced a sweet taste and pressed for government certification. However, in September 1980, an FDA public board of inquiry concluded aspartame should not be approved pending further investigations of brain tumors in animals. It has been reported that later that year, during a sales meeting, the then CEO of G. D. Searle, future secretary of defense Donald Rumsfeld, stated that he planned to use his political pull in Washington to get aspartame approved.

The story is as follows: On the day after President Ronald Reagan's inauguration in 1981, Rumsfeld, as a member of the new administration, was instrumental in the appointment of Dr. Arthur Hull Hayes as FDA commissioner. Hayes, who had previously served in the U.S. Army's Chemical Weapons Division, initially approved aspartame as a sweetener powder, despite the objections of three of six FDA scientists responsible for researching it. Drs. Robert Condon, Satya Dubey, and Douglas Park argued against approval, stating on the record that Searle safety tests were unreliable and not adequate to determine the safety of aspartame.

In 1983, just before he left his FDA position for a public relations job with Burson-Marsteller, the chief public relations firm for both Monsanto and Searle, Hayes approved aspartame for all carbonated beverages.

In 1985, Searle was purchased by multinational conglomerate

Monsanto, which established the NutraSweet Company. NutraSweet was bought by the European firm J.W. Childs Associates in 2000.

Aspartame's connection to formaldehyde, a deadly neurotoxin, has raised concerns among many regulators. A 1998 *Life Sciences* study concluded that "aspartame consumption may constitute a hazard because of its contribution to the formation of formaldehyde adducts [two or more compounds combined]." It has been established that aspartame releases free methanol that breaks down into formic acid and formaldehyde in the human body when subjected to temperatures of more than eighty-six degrees Fahrenheit. Normal body temperature is 98.6 degrees. One quart of an aspartame-added beverage is estimated to contain about fifty-six milligrams of methanol.

Today, however, the FDA and European Food Safety Authority maintain that aspartame is safe at the levels currently used in food and drinks. Yet aspartame accounts for more than 75 percent of adverse reactions to food additives reported to the FDA, according to published reports. These include seizures and death. The more than ninety other documented symptoms caused by aspartame listed in a FDA report include headaches/migraines, dizziness, seizures, nausea, numbness, muscle spasms, weight gain, rashes, depression, fatigue, irritability, tachycardia, insomnia, vision problems, hearing loss, heart palpitations, breathing difficulties, anxiety attacks, slurred speech, loss of taste, tinnitus, vertigo, memory loss, and joint pain. According to researchers and physicians studying the adverse effects of aspartame, the following chronic illnesses can be triggered or worsened by ingesting aspartame: brain tumors, multiple sclerosis, epilepsy, chronic fatigue syndrome, Parkinson's disease, Alzheimer's, mental retardation, lymphoma, birth defects, fibromyalgia, and diabetes. Despite this laundry list of related ailments, the sweetener, which is sold under a variety of names including NutraSweet, Equal, E951, Benevia, and Canderel, is ubiquitous. Zachary Shahan, director of CleanTechnica, a popular website devoted to science news and conservation, estimates more than 250 million people, or about two-thirds of the U.S. population, consume aspartame in its various forms.

Academics who have studied the true effects of aspartame find it a thankless job. In 1991, Dr. Janet Starr Hull, an OSHA-certified environmental hazardous waste emergency response specialist and toxicologist, was diagnosed with incurable Grave's disease (a defect in the immunization system that leads to hyperthyroidism) only to learn through her own research that she had been poisoned by aspartame. She stated: "Many scientists at prestigious American universities will tell you they cannot get grants for continued research on aspartame or Splenda, or their department heads have been told to drop all discussions on the topic. Some will say aspartame research isn't worth the effort because they cannot get published in American scientific journals. Others claim the research centers constructed by the large corporations, such as Duke University's Searle Research Center, were designed with managed research as a construction proviso."

In addition to ethical questions concerning drug corporations funding academic studies and research centers, eyebrows have been raised over corporate financing of NutraSweet researchers.

For example, former FDA investigator Arthur M. Evangelista noted that Susan Schiffman, named to head a Searle-funded Duke University Medical School study into NutraSweet's link to headaches, is a former General Foods and Searle consultant. Her research is under the auspices of the office of university vice president William Anylan, a former Searle director. Schiffman said she took no salary for the research and that Anylan played no part in Searle's pledge to finance the study, expected to cost hundreds of thousands of dollars. Other academics mentioned by Evangelista include Dr. David Hunninghake of the University of Minnesota, who was selected to head a Searle-designed study of aspartame's effect on the liver by former Searle research director Daniel Azarnoff, formerly Hunninghake's mentor at the University of Kansas.

It also has been reported that Dr. Lewis Stegink, a pediatrics professor at the University of Iowa whose studies since the early 1970s invariably indicated aspartame's safety, at one point had received more than $1.3 million dollars in research grants and gifts (including lab equipment) from the NutraSweet company, according to university records.

A longtime research collaborator of Stegnik's, Dr. Jack Filer serves as executive director of the International Life Sciences Institute (ILSI), a Washington, D.C., foundation funding aspartame research, among other things. Filer claims he sees no conflict in his dual roles as ILSI's executive director and an aspartame researcher, but declined to disclose his ILSI consulting fees. Filer has argued that health problems blamed on aspartame may simply be "water load" on the brain caused by the overconsumption of liquids.

Such apparent conflicts of interest seem to have prevented inquiry that is desperately needed. While not focusing specifically on aspartame or other artificial sweeteners, a 2014 University of Iowa study of diet drinks found that otherwise healthy women who were regular consumers of diet drinks were significantly more likely to die from a heart attack or cardiovascular disease. Nearly sixty thousand women participated in the study, which found those who consumed two or more diet drinks a day were 30 percent more likely to have a cardiovascular event and 50 percent more likely to die from heart disease than women who never or rarely drank diet drinks.

"This is one of the largest studies on this topic, and our findings are consistent with some previous data, especially those linking diet drinks to the metabolic syndrome," said Dr. Ankur Vyas, lead investigator of the study, who cautioned that further research is required. "It's too soon to tell people to change their behavior based on this study; however, based on these and other findings we have a responsibility to do more research to see what is going on and further define the relationship, if one truly exists. This could have major public health implications."

Older readers might recall that for decades the tobacco industry denied any correlation between disease and smoking. Doctors and celebrities would recommend that people smoke to appear sophisticated or to calm their nerves. Could today's doctors recommending diet drinks be equally misinformed?

In an article published by Inquisitr.com, Terri LaPoint noted that Searle, the manufacturer of aspartame, began developing a drug to combat memory loss shortly after the FDA approved aspartame for use in carbonated drinks. "Aspartame is a neurotoxin. Even ants have sense enough to avoid it," she explained. "Yet diet drinks add this neurotoxic chemical as its sweetener, and they promote it as a heath food to a public that naively puts its trust in the experts. Then the manufacturers stand ready to offer you drugs to help you with your symptoms that they don't tell you are directly related to your diet sodas. It's a win-win situation for them, with the consumer as the loser. You don't lose weight. You lose health."

Some physicians have even linked aspartame to rising rates of amyotrophic lateral sclerosis (ALS), commonly known as Lou Gehrig's disease. A study of U.S. troops deployed in the Persian Gulf during the First Gulf War found that they had twice the risk of developing ALS as compared to military personnel who were not deployed. Dr. Betty Martini points out that these troops were consuming lots of diet soda, which, in conjunction with other environmental stressors, may be connected to the disease. "In the Persian Gulf at that time the troops were consuming lots of diet pop cooking in the 120-degree Arabian sun for as long as eight weeks at a time, according to vets I personally interviewed in Huntsville, Alabama," reported Dr. Martini, a member of Mission Possible World Health International.

Aspartame has generated such bad press that one large manufacturer has distanced itself from the chemical. The Japanese food and additive corporation Ajinomoto Kabushiki-kaisha or simply Ajinomoto Co. Inc. was originally a primary maker of MSG. Today it is the world's largest producer of aspartame, with a 40 percent global market share. In 2009, Ajinomoto began using the name Amino-Sweet to avoid the controversy of the name aspartame.

OTHER SOFT DRINK HAZARDS

AS THOUGH ASPARTAME WERE NOT WORRY ENOUGH, OTHER SODA ingredients have begun to come under intense scrutiny. A UK study in 2007 revealed a salt derived from benzoic acid and routinely used as a preservative by the $160 billion soft-drink industry changes to cancer-causing benzene when mixed with vitamin C in liquids.

Peter Piper, a professor of molecular biology and biotechnology at Sheffield University, found that sodium benzoate damaged the mitochondrial DNA of yeast cells. Mitochondria are tiny free-floating elements within each cell that convert the energy from food into a form the cells can use. "These chemicals have the ability to cause severe damage to DNA in the mitochondria to the point that they totally inactivate it: they knock it out altogether . . . there is a whole array of diseases that are now being tied to damage to this DNA—Parkinson's and quite a lot of neuro-degenerative diseases, but above all the whole process of aging," warned Piper. Although the strength and dosage of sodium benzoate needed to produce cancers requires further study, many health authorities advise avoiding all soft drinks.

Other studies have tentatively associated sodium benzoate with attention deficit disorder (ADD) but these studies focused on food dyes, also accused of creating health hazards, as well.

Benzene is a known carcinogen that usually is found in cigarette smoke, automotive exhaust fumes, industrial waste, and around service stations. It is sometimes found in contaminated food and water, including some soft drinks. Furthermore, the long-term effects of sodium benzoate poisoning are simply not known.

If the chemicals in drinks don't get you, the containers might. Whether diet or regular, all soft-drink cans are coated with the endocrine disruptor bisphenol A (BPA), which has been linked to everything from heart disease to obesity to reproductive problems.

Recent research indicates that BPA is a potent mimicker of the estrogen hormone, and urinary concentrations of BPA have been linked to obesity, especially in children. Dozens of studies have linked BPA with prostate cancer, infertility, asthma, heart disease, and neurodevelopmental disorders.

Predictably, in July 2014, the FDA reported that its latest studies indicated BPA is "safe at the current levels occurring in foods." However, the European Food Safety Authority (EFSA) continues to recognize some uncertainty about this "safe" synthetic compound.

In early 2015, scientists at the University of Calgary reported concern over bisphenol S (BPS), thought to be a less harmful version of BPA. It is found in many products carrying "BPA-free" labels. The study, published in the *Proceedings of the National Academy of Sciences,* found low doses of BPA and BPS to cause underdevelopment and hyperactivity in zebra fish, which share 80 percent of their genes with humans.

Many plastics also contain dioxins, which have been shown to cause cancer, especially breast cancer. These chemicals from plastics find their way into our bodies in a number of ways. As Dr. Edward Fujimoto, director of the Center for Health Promotion at Castle Medical Center in Hawaii, warns, freezing plastic bottles with water in them can release dioxins, as can heating fatty foods in plastic containers in microwave ovens. The combination of fat, high heat, and plastics releases dioxin into the food and ultimately into the cells of the body. Frozen TV dinners, instant ramen and soups, and such should be removed from their container before heating. Fujimoto also notes that the presence of dioxins is one reason some fast-food chains have moved away from foam containers, using paper instead. Fujimoto recommends using glass, such as CorningWare, Pyrex, or ceramic containers, for heating food.

Yet the FDA, while not oblivious to the plastics problem, has willfully disregarded the extent of the risk. The agency admits that the substances used to make plastics may leach into foods, yet it claims that the levels in foods are within the margin of safety. The FDA website adds it has "seen no evidence that plastic containers or films contain dioxins and knows of no reason why they would," apparently disregarding clear evidence to the contrary.

Though the FDA continues to play dumb on these safety issues, researchers' concerns have begun trickling down to consumers, whose wariness over the health dangers of soft drinks is beginning to show in soft-drink sales. Industry newsletter *Beverage Digest* reported in 2014 that sales fell 3 percent the previous year, with the number of cases sold reaching a nearly two-decade low. Industry analysts attributed the decline to growing public awareness of the health concerns over artificial sweeteners.

Yet the industry largely continues to resist improving safety and regulation. Americans concerned about their health must take charge themselves. Eliminating use of plastics containing BPA and dioxins is a good first step. Several natural sweeteners are also likely to be significantly less harmful than their artificial counterparts. Monkfruit extract (also known as *luo han guo* or *lo han kuo*) is two hundred times sweeter than sugar but has a licorice-like aftertaste, and monatin, which is made from a South African shrub, is said to be three thousand times sweeter than sugar. Neither monatin nor monkfruit are known to contain harmful chemicals, though it must be noted that neither has undergone significant scientific tests.

Another popular alternative to artificial sweeteners is the easy-to-grow natural herb stevia, used as a sweetener in Brazil for centuries. Some studies indicate stevia actually is an insulin sensitizer that can aid in reducing blood glucose and even improve memory. Health authorities, noting the lack of long-term studies on the regular use of stevia compared to the known dangers of Splenda, suggest users practice moderation in regard to both.

Viewing the vast amount of literature and research pointing to potential health hazards in both the use and overuse of artificial sweeteners, the health-conscious reader should consider giving up all carbonated soft drinks. But since this is unlikely in this sweetness-addicted nation, one should at least make a modest effort to learn which substances should be eliminated and which should be limited in consumption.

Only one thing is for certain in considering sweeteners: do not allow either the government or the profit-seeking food and drink industry to make decisions for you.

CHAPTER 12

GMOs

As if chemicals, pollution, pesticides, and poisons weren't dangerous enough, the past two decades have also seen the emergence of genetically modified organisms (GMOs), today the subject of a growing controversy.

GMOs, which now appear in up to 80 percent of processed food, are plants or animals whose cells have been inserted with a gene from an unrelated species, sometimes even a virus or bacterium, in order to create a specific characteristic, often one that resists insects or blight. Although GMO-heavy foods may have a longer shelf life, bringing in greater profits for the manufacturer, the health costs are deeply troubling. These experimental combinations of genes from different species do not occur in nature or traditional crossbreeding, and no genetically engineered animal has been approved for human consumption in the U.S.

Early in 2015, Russian president Vladimir Putin signed a law that mandated the labeling of food products containing GMOs. The new law was supported by Prime Minister Medvedev, who stated, "The government will not poison their citizens."

Russia joined more than sixty-four countries worldwide, including fifteen nations in the European Union, Japan, Australia, Brazil, and even China, that now require labeling of GMO foods. Many have placed restrictions or outright bans on the production and sale of GMOs. But in the U.S., money talks. The federal government has regularly approved GMO products based on studies conducted by the same corporations that profit from their sale.

As the evidence against GMOs continues to pile up, it is becoming obvious that the federal government is incapable of providing adequate safety research for the public. Apparently biotech lobbying is so pervasive that the government cannot even impose clear labeling standards.

In early 2015, a book by Steven Druker entitled *Altered Genes, Twisted Truth—How the Venture to Genetically Engineer Our Food Has Subverted Science, Corrupted Government, and Systematically Deceived the Public* was published to acclaim by scientists knowledgeable on the GMO issue. Druker's work revealed how for more than three decades, hundreds of eminent biologists and revered institutions have systematically contorted the truth about GMO foods and concealed its unique risks.

In the book's foreword, the famous award-winning English anthropologist Jane Goodall attacked the idea that genetically engineered foods are no different from natural crops and therefore safe. She noted "how amazingly successful the biotech lobby has been— and the extent to which the general public and government decision-makers have been hoodwinked by the clever and methodical twisting of the facts and the propagation of many myths. Moreover, it appears that a number of respected scientific institutions, as well as many eminent scientists, were complicit in this relentless spreading of disinformation."

Goodall labeled the push for GMO foods "the biggest scientific fraud of our age," and added "the key step in the commercialization of GE foods occurred through the unbelievably poor judgment—if not downright corruption—of the US Food and Drug Administration (the FDA) . . . it apparently ignored (and covered up) the concerns of its own scientists and then violated a federal statute and its own regulations by permitting GE foods to be marketed without any testing whatsoever."

According to economist Dr. Paul Craig Roberts, who served

as assistant secretary of the treasury in the Reagan administration, Dick Cheney used his two terms as vice president to fill environmental agencies, including the FDA, with corporate-friendly executives.

Jeffrey M. Smith, executive director of the Institute for Responsible Technology, proclaimed, "In the critical arena of food safety research, the biotech industry is without accountability, standards, or peer-review. They've got bad science down to a science."

Despite growing public opposition to the use of GMOs, citizens in some states have been unable to pass legislation simply requiring the labeling of GMO products because the chemical and junk-food industries have spent tens of millions of dollars to make sure Americans are kept in the dark. According to the Environmental Working Group, the Grocery Manufacturers Association (GMA), Coca-Cola, PepsiCo, Monsanto, along with the chemical firm DuPont and others spent more than $27 million just in the first half of 2014 to lobby against GMO labeling.

In spite of the money spent and narrow defeats at the polls, the movement to require GMO labeling continued to gain strength and momentum, with fourteen states considering laws requiring such labels.

In May 2014, Vermont became the first state to require GMO labels. The new law will take effect in mid-2016 if it survives legal challenges. Maine and Connecticut passed laws before Vermont, but those measures don't take effect unless neighboring states follow suit.

A 2014 California vote on Proposition 37, which would have required GMO labeling, was narrowly defeated due to an estimated $46 million spent to oppose the measure by food giants including Kellogg's, General Mills, PepsiCo, and Monsanto. In Oregon, a statewide vote on labeling (Measure 92) came within five-hundredths of one percent of winning.

Dr. David Bronner, president of Dr. Bronner's Magic Soaps and a blogger, stated, "The bottom line is, the GMO labeling movement is on fire and surging. We . . . are as likely to achieve victory through the market by 2016, as we are unleashing and fueling major cultural and market drivers and expect more and more food companies to flip and accept mandatory labeling just as they did in Europe."

The federal government has also shown that it is able to override local opposition to GMOs. Kaua'i County in Hawaii in late 2013 passed an ordinance requiring disclosure of some pesticide use as well as GMO crop cultivation, with some restrictions on crops near schools and nursing homes. But in August 2014, U.S. Magistrate Judge Barry Curren overturned the ordinance, saying it was pre-empted by other laws and thus invalid. Curren's action was viewed as a victory by seed and chemical companies in a battle over modern agricultural techniques. DuPont, Syngenta AG, Dow Chemical, and BASF were among those who challenged the ordinance and jointly expressed pleasure with the ruling.

Gary Hooser, a Kaua'i council member who cointroduced the ordinance, said the controversy is far from over. "One opinion from one federal magistrate does not settle the issue," he said.

The GMO controversy is a morass, with well-meaning and passionate advocates on both sides. Bringing suspicion on the pro-GMO proponents is the view that most of them are in one way or another funded by the very corporations that profit from gene modification.

Critics offer more emotional arguments pointing to a variety of complaints, whether real or imagined. Anecdotes of GMO harm range from birth defects in Danish pigs fed GMO soy and dying sheep and goats in India to soy allergies skyrocketing in the UK and the multiple horror stories of lab rats suffering from organ lesions, infertility, altered enzyme levels, and liver and pancreas problems.

Books such as *The World According to Monsanto* and *Seeds of Destruction* paint the giant chemical corporations as mass murderers. They claim that since 1901 the firms have produced saccharin, PBCs, BPA, BPS, glyphosates, fungicides, pesticides, herbicides, insecticides, Agent Orange, napalm, DDT, neotame, aspartame, chemical and petrochemical fertilizers, heavy metals, and GMOs, all of which have collectively killed more humans globally than all global terrorism combined over the past two millennia.

The American Academy of Environmental Medicine (AAEM), an international group of physicians and associated professionals dedicated to the study of environmental illnesses, has reported, "Several animal studies indicate serious health risks associated with

GM food." These risks include infertility, immune system problems, accelerated aging, faulty insulin regulation, and changes in major organs and the gastrointestinal system. The AAEM has asked doctors to advise patients to avoid GMO foods.

One noteworthy complaint against GMOs is that, unlike drug safety evaluation, there have been no human trials of GMO foods. The Institute for Responsible Technology (IRT), which was founded in 2003 and is now active in forty countries, noted on its website the only published human feeding experiment revealed that the genetic material inserted into GM soy can transfer into the bacteria living inside our intestines and continue to function. This means that long after we stop eating GMO foods with an antibiotic gene, we may still have this gene inside us, creating antibiotic-resistant superdiseases. If the gene that creates Bt toxin in GM corn were to transfer, it might turn our intestinal bacteria into living pesticide factories.

One good illustration of the ongoing controversy over GMOs can be found in the story of a study conducted by Gilles-Éric Séralini, a professor of molecular biology at the University of Caen in France, and the founder and president of the scientific advisory board of the Committee of Research and Independent Information on Genetic Engineering (CRIIGEN). He is also a member of the EU's Commission for Biotechnology Reevaluation, created in 2008.

In 2012, Séralini and colleagues published a paper entitled "Long-term Toxicity of a Roundup Herbicide and a Roundup-Tolerant Genetically Modified Maize" in the journal *Food and Chemical Toxicology*. The research, based on a two-year study of Monsanto's genetically modified maize NK603, engineered to be resistant to the herbicide Roundup, reported that mice who were fed the GM corn had an increase in both body and liver weight.

"For the first time in the world, we've proven that GMOs are neither sufficiently healthy nor proper to be commercialized . . . Each time, for all three GMOs, the kidneys and liver, which are the main organs that react to a chemical food poisoning, had problems," proclaimed Séralini.

This study prompted rapid and vitriolic responses both in Europe and America. Despite support for Séralini 's article in an open letter signed by about 130 scientists, scholars, and activists published in *Independent Science News* and from proponents of California's GMO labeling proposition, many mainstream organizations viciously attacked the paper.

The European Food Safety Authority (EFSA) accused the study of being inadequately designed, analyzed, and reported and being "of insufficient scientific quality for safety assessments."

The study was accused of "numerous issues relating to [its] design and methodology," including using too small a sample of rats yet it was pointed out that it was the same as toxicity studies reported by Monsanto, one difference being that the Séralini research was conducted for more than two years rather than the 90 days of the Monsanto study which affirmed GMO safety. Since Séralini's study criteria seems to have matched or exceeded at least one of the Monsanto protocols, his results should be considered at least as valid as Monsanto's. Critics say Séralini's findings prove that the Monsanto short-term safety studies are flawed.

Due to the outcry over Séralini's study, the editors of *Food and Chemical Toxicology* retracted his paper in 2013 despite Séralini's objections and despite the fact that no charges of fraud or misrepresentation were lodged. In June 2014, the paper was republished in the journal *Environmental Sciences Europe*.

But accusations that GMO safety has been left up to the same corporations that profit from gene modification continue to grow, as does public concern over the long-term safety of GMOs.

Meanwhile, in what some see as a bait-and-switch scheme, some GMO companies are now using names such as Biologics, Natural Identicals, and Biologics Identicals to mask the use of GMOs from the public.

Furthermore, the Union of Concerned Scientists, following a review of GMO farming in the U.S., determined that GMOs do not increase crop yields.

GMO critics have understandably focused on the harmful effects of GMOs on the human body, but we should also look at their environmental effects. GMOs, used commonly on crops such as corn, canola, soybeans, and cotton, were initially hailed as a means of reducing pesticide use. This claim has since been called into question, as overreliance on these crops has led to the emergence of "super-weeds," which are more resistant to herbicides and require increased spraying.

The Non-GMO Project, a nonprofit that aims to achieve a GMO-free food supply, explained, "Over 80 percent of all GMOs grown worldwide are engineered for herbicide tolerance. As a result, use of toxic herbicides like Roundup has increased fifteen times since GMOs were introduced. GMO crops are also responsible for the emergence of 'super weeds' and 'super bugs': which can only be killed with ever more toxic poisons like 2,4-D [2,4-Dichlorophenoxyacetic acid, a major ingredient in Agent Orange]. GMOs are a direct extension of chemical agriculture, and are developed and sold by the world's biggest chemical companies. The long-term impacts of GMOs are unknown, and once released into the environment, these novel organisms cannot be recalled."

In 2014, environmentalists urged the Environmental Protection Agency (EPA) not to approve a newly developed herbicide called Enlist Duo, produced by Dow Agrosciences. This chemical, which combines glyphosate with 2,4-D in an effort to intensify both, may lead to even more herbicide-resistant weeds, according to critics. Enlist Duo was developed in response to weeds resistant to glyphosate found in twenty-seven states.

It seems that in addition to the health hazards of glyphosate, this widely used herbicide has failed to fulfill its promise to rid crops of weeds.

Jay Feldman, executive director of the environmental group Beyond Pesticides, noted that the EPA rejected a request by the state of Texas to allow the emergency (meaning nonregistered) use of the herbicide propazine on three million acres of cotton due to the failure of glyphosate to kill some weeds as well as concerns over water contamination. Feldman said approval of Enlist Duo would only be throwing a "life preserver" to growers and perpetuate a "fatally flawed technology."

Feldman and other GMO critics claim unrestricted use of genetically modified herbicides could conceivably create a disaster in the life-essential food chain. "The problem of weed and insect resistance to pesticides was predictable when herbicide-tolerant and pesticide-incorporated plants were introduced," said Feldman. "The promise of genetic engineering for these characteristics has failed as a sustainable practice; first, with increasing glyphosate use—and now the collapse of the system."

And the footprint of GMOs goes far beyond pesticides and the effects on individual crops. A recent study even indicated that GMO-engineered plants may be partly responsible for the release of carbon into the atmosphere, causing an increasingly warm environment.

For some time, carbon emissions have been attributed largely to the human population, polluting SUVs, and methane from flatulent cows. In 2014, a paper published in Yale University's *Yale Environment 360* entitled "Soil as Carbon Storehouse: New Weapon in Climate Fight?" suggested that GMOs and attendant toxic pesticides and herbicides may also be a major cause of carbon release.

According to scientists, more carbon resides in soil than in the atmosphere and all plant life combined. One report stated there are 2,500 billion tons of carbon in soil, compared with only 800 billion tons in the atmosphere and 560 billion tons in plant and animal life.

When soil is damaged as a result of unsustainable agriculture practices, like planting millions of acres of GMO soy and corn, and the regular spraying with toxic pesticides and herbicides, large amounts of carbon are released into the atmosphere. In contrast to healthy soil, which naturally sequesters carbon and preserves it for the benefit of both humans and the environment, aggressive farming with the aid of GMOs leads to the release of greater amounts of carbon. Interestingly, one notable way of restoring soil is the planting of cannabis or hemp, which absorbs more CO_2 than any other tree, shrub, or plant known to man. But growing cannabis remains illegal in the United States.

Rattan Lal, director of Ohio State University's Carbon Management and Sequestration Center, reported that the world's cultivated soils have lost between 50 and 70 percent of their original carbon stock, much of which has oxidized upon exposure to air to become carbon dioxide (CO_2). Up until now, the debate on climate has centered on reducing carbon emissions. "Reducing emissions is crucial," Lal remarks, "but soil carbon sequestration needs to be part of the picture as well." Lal adds that restoring degraded and eroded lands, as well as avoiding deforestation and the farming of peat lands, should be top priorities.

The destructive effects of corporate agriculture may have been felt on a grand scale—climate change—but also in many smaller ways. A troubling mystery was solved recently when the use of insecticides was linked to the strange disappearance of honeybees, a phenomenon widely reported in the media. The worldwide loss of honeybees has caused concern because about three-quarters of the world's food crops require pollination.

Writing in the *Bulletin of Insectology,* researchers reported finding a direct link between the use of neonicotinoids, the most widely used class of insecticides, and honeybee colony collapse disorder (CCD). "We demonstrated that neonicotinoids are highly likely to be responsible for triggering 'colony collapse disorder' in honeybee hives that were healthy prior to the arrival of winter," reported Chensheng Lu of the Harvard School of Public Health and lead researcher of the study.

In a 2014 interview with the *Boston Globe,* Sheldon Krimsky, a professor at Tufts University, the head of the Council for Responsible Genetics, and author of *The GMO Deception,* said that true science demands caution, especially when changing the genetic makeup of our food. Krimsky explained, "The problem with GMOs goes back to 1992 after the Quayle Commission (named for then-vice president Dan Quayle) issued guidelines for biotechnology. That report advised the FDA that 'you didn't need to test any of these products.' They simply told industry if you see any problems, let us know . . . You cannot predict what's going to happen to an organism if you put in a foreign gene. It could interfere with other genes, it could over-express some things and under-express other things. You cannot make predictions without testing them . . . Genes do more than one thing. If you think of the genome as an ecosystem rather than a LEGO system, it gives you a different idea of what the possibilities are. We have to test in order to understand what the foreign gene is going to do to the organism."

Krimsky said creating GMOs is not the same as breeding hybrids, which come from the same species. Genetic modification involves putting foreign genes from separate species into an organism. He said comparing hybrids to GMOs would be "like saying let's put a few animal genes into the gamete of human beings and assume that it's no different than if we just threw in some genes from another human being."

Citing a peer-reviewed French study that suggested GMOs caused cancer in lab rats, Krimsky defended the work by explaining, "When you're looking at risks for a product or technology, it is rational to look at worst-case scenarios. If you're testing the safety of a new airplane, you want to test it at the limits, not in safe flying conditions . . . Whenever you're looking at the risk of a product, a single negative result is more important than ninety-nine positive results, especially when a substantial number of those positive results are funded by the agribusiness industry. We've had products on the market for fifty years: PCBs, asbestos, tobacco, DDT. Early on people said they were safe."

Krimsky, noting his family tries to buy only organic food, said he would feel comfortable with GMO products only if there were "independent studies asking the right questions and seeking experiments looking for the most vulnerable cases."

Money and politics have prevented such studies from being conducted. In mid-2014, the UK's *Daily Mail* reported on emails documenting collusion between the GMO industry and the British government. The emails revealed a broad and deliberate strategy designed to thwart European regulations on GMO crops.

They asked industry lobbyists for "eye-catching themes" on GMOs to present to government officials and even spoke of the creation of a blacklist of journalists who wrote stories on potential GMO hazards.

Dr. Helen Wallace, director of GeneWatch, an environmental group that obtained the emails through a Freedom of Information request, says the public would be shocked at the level of collaboration between the pro-GMO Agricultural and Biotechnology Council (ABC) and British ministers that "shifted government policy to support GM, despite clear opposition among consumers." The ABC is supported by GMO manufacturers such as Monsanto, Syngenta, and Bayer CropScience. "These documents expose government collusion with the GMO industry to agree on PR messages and blacklist critical journalists," explained Wallace.

British journalist Sean Poulter reported that the email exchanges showed that UK environment secretary Owen Paterson pushed the EU to allow GMO crops in Britain even though they were banned elsewhere, while Science Minister David Willetts supported a pro-GM "Agri-tech" strategy to develop new crops with public money.

"Such support from two key government figures represents a coup for the GM industry," reported Poulter, who noted the email exchanges often coincided with government announcements giving support for GMOs despite consumer opposition.

Despite the industry's efforts to suppress GMO labeling and public awareness, it's clear that most Americans want these products clearly marked. A 2012 Mellman Group poll reported 91 percent of consumers wanted GMO labeling, while a 2008 CBS/*New York Times* poll indicated 53 percent of respondents said they would not buy food that has been genetically modified. A December 2014 Associated Press–GfK poll showed 66 percent of Americans support GMO labeling whether or not they consume such products. According to former FDA adviser Marion Nestle, the battle over GMO labeling boils down to one basic fear by GMO companies. "They didn't want it labeled because they were terrified that if it were labeled, nobody would buy it," she says.

As both public awareness and concern over GMOs increases, some companies have begun moving away from them. General Mills announced in 2014 that its iconic cereal brand Cheerios will no longer contain GMO ingredients, though the company continues to fund the fight against the labeling of GMO products. In mid-2014, General Mills purchased Annie's, the popular maker of organic mac and cheese and other snack products, for $820 million. Some observers suggested the move could signal GM shifting its product line into purer foods, but others wondered whether the giant corporation was instead planning to add GMO ingredients to Annie's products. Herein lies the issue with GMOs: it's often difficult to discern which products contain them, and food corporations can profit by being coy.

In recent years, a few forward-thinking grocery stores and restaurant chains have made efforts to label GMO foods so that consumers can at least know which products are GMO and which are not. The Chipotle restaurant chain announced it would be removing GMO ingredients from its menu. Whole Foods Market said it will require the labeling of all GMO products at its U.S. and Canadian stores by 2018. But such measures as public policy have been voted down after Monsanto and GMO-friendly corporations poured millions into campaigns to prevent such labeling.

Some question whether GMO labeling would even help. In mid-2014, the House Agriculture Committee's Subcommittee on Horticulture, Research, Biotechnology and Foreign Agriculture heard from several GMO-friendly academics who essentially argued that Americans were too ignorant to read and understand GMO labeling. David Just, a Cornell University professor and codirector of the Cornell Center for Behavioral Economics in Child Nutrition Programs, told the subcommittee, "It is ignorance of the product, and it's a general skepticism of anything they eat that is too processed or treated in some way that they don't quite understand. Even using long scientific-sounding words make it sound like it's been grown in a test tube, and people get scared of it." Another academic, Professor Calestous Juma of Harvard's Kennedy School, agreed, adding that misinformed voters have cowed political leaders, especially in the EU, into placing restrictions on the use of GMOs.

Not called to testify before the House subcommittee was Jerry Greenfield, cofounder of Ben & Jerry's ice cream, who disputes the argument that GMO labeling would frighten consumers. "This idea that consumers will be scared away—the label will be a very simple thing, a few words on a container saying something like 'may be produced with genetic engineering.' It's not scary," Greenfield says.

Many people, including those in favor of GMO products, view the consumption of GMO food as an individual choice, a freedom, but one that should involve informed consent, especially given the controversy the subject has engendered. Everyone should be informed as to which products are GMO and which are not. Andy Kimbrell, executive director of the Center for Food Safety, sees labeling as the only way health professionals may be able to trace problems. "For example, if you're a mother and you're giving your child soy formula and that child has a toxic or allergic reaction, the only way you'll know if that's a genetically engineered soy formula is if it's labeled," he explained.

The human body might be able to cope individually with many of these harmful chemicals and even GMOs. What worries health advocates and nutritionists is the cumulative effect over the long term.

Another concern is the corporations and other "middlemen" who take the better portion of the food dollar. Tracie McMillan, a senior fellow at the Schuster Center for Investigative Journalism, explained, "When we buy food, we think we are paying the farmer. This is true in a very basic economic sense: some portion of what we spend at the store does trickle back down to the hands that worked the land. Understandably, we think that if food costs more, it must be because the farmer is getting more for it . . . The problem is, that is almost entirely untrue."

She estimates that only about sixteen cents of every food dollar goes to the farmer. The remaining eighty-four cents goes for what is called marketing. This refers not only to commercials and advertising, but the entire chain that makes sure food gets from the farm to the dinner table. It includes the trains and trucks and drivers who move the food from farm to processing plant or warehouse, the mill or the factory where food is processed, and the cost of storing it until it is sold. The dollar also pays the wholesalers and retailers, the grocers, the restaurant cooks who prepare it for us when we eat out, the satellites and databases used to track shipments, and the workers, forklifts, warehouse, and refrigeration at the grocery store.

Still, the bulk of the food dollar goes to one of the ten giant multinational corporations that control much of the food supply of the developed world: Nestlé, Coca-Cola, PepsiCo, General Mills, Kellogg's, Mars, Mondelçz, Associated British Foods, Danone, and Unilever.

These giant food corporations, which have gained dominant control over our food supply, are the ones most opposed to food labeling. Most families now eat meals that come from supermarket chains or supercenters (stores like Walmart and Target). Walmart, which sells about one-quarter of all the food purchased in the U.S., is now the largest retailer in history, according to *Forbes.*

Some of the more familiar products made by these firms include the Skinny Cow owned by Nestlé, Tropicana orange juice owned by PepsiCo, Ben & Jerry's ice cream by Unilever, Oreo cookies by Mondelçz International, Dasani water by Coca-Cola, Twinings tea by Associated British Foods, Dannon yogurt by Danone, Old El Paso Mexican foods by General Mills, M&M's candy by Mars, and Pop-Tarts along with several cereals owned by Kellogg's.

"Those ten companies . . . are now the biggest food and beverage companies in the world," noted Oxfam America, an international organization working to eradicate world poverty. "Together, they generate revenues of more than $1.1 billion a day. They also employ millions of people in poor countries, directly and indirectly, to grow and produce their products. Because of their global reach and influence, these companies could play a big role in reducing poverty, hunger, and inequality. But right now, they're not doing enough." A remarkable chart of the many food companies and brand names owned or controlled by the ten mega-corporations may be found at the Oxfam website listed in the notes section.

"With a mere ten companies controlling the world's food supplies it should give a person pause regarding not only what we are actually eating, but what makes up what we put into our bodies," wrote Rory Hall in the *Daily Coin.* "When we think of food, our minds automatically revert to nutritional items that sustain our lives and keep us healthy. Some of you may think of vegetables or grains or a well-marinated steak. Whatever pops into your mind, rest assured it probably is not on [the Oxfam chart]. Well, it may be on the [chart], just not how you may be thinking of it in terms of nutritional value, wholesomeness, and advancing or maintaining your overall health."

"If someone can help me to understand that this is not a conspiracy, I would really be appreciative. Otherwise, it sure looks like a controlled plan working against humanity," concluded Hall.

The ten dominant food corporations are owned by globalists or controlled by their banks. This presents a scary scenario of a future in which whole populations are held under the thumb through the control of food and water. Today a handful of globalist-controlled banks manipulate the world's financial system and they are now working hard to control food and water, both essential to life. After all, whoever controls the food controls the people.

Genetic engineering and such products as Monsanto's "Terminator" seeds, which cannot reproduce, could conceivably give that company proprietary control over the food crops.

Glyphosate-based Roundup with its complex of "inert" ingredients, touted as a benign substitute for poisonous dioxin herbicides, is insidious, as it may limit access to essential nutrients even as it destroys everything in its path except those plants genetically modified to resist it.

Could such chemical herbicides along with GMOs explain why the World Health Organization has ranked the U.S. at the bottom of a list of seventeen developed nations in overall health? Could the top-down pressure to ruin soil, destroy natural plants, and adversely impact the public health through GMOs be part of a global conspiracy of elitists who make no secret of wanting population reduction?

It is not just food that is costing the public both money and health, but also the deadly water we drink.

CHAPTER 13

DEADLY WATER

DO YOU KNOW WHAT IS IN THE WATER THAT YOU DRINK? DO YOU really want to know?

The comedian W. C. Fields once quipped that he never drank water because fish fornicate in it. But toxic water is no joke. According to the National Institute of Environmental Health Sciences, nearly two billion people worldwide drink harmful contaminated water. This water is polluted by fertilizers and pesticides from agricultural runoff; sewage and food processing waste; lead, mercury, and other heavy metals; chemical wastes from industrial discharges; and chemical contamination from hazardous waste sites.

Most Americans take their water for granted, expecting government to ensure its purity and cleanliness. However, public water supplies may be behind many of the health problems facing the nation. Researchers from the U.S. Geological Survey (USGS) and the U.S. Environmental Protection Agency (EPA) have found that an astounding one-third of U.S. water systems contain traces of at least eighteen unregulated and potentially hazardous contaminants, many of which have been shown to cause endocrine disruption and cancer. Is this the result of inattention on the part of water authorities or could it be part of a plan to sicken and eliminate portions of the population?

A wide range of disturbing materials have been found in American water. In a nationwide survey of twenty-five water utilities, scientists found traces of the herbicide metolachlor, a pesticide commonly applied to conventional corn, soy, cotton, safflower, potato, and other crops, as well as the heavy metal strontium, which is linked to bone problems. Other chemicals identified include so-called perfluorinated compounds like perfluorooctanoic acid (PFOA), which numerous scientific studies have found can cause thyroid disease and various types of cancer.

Of further concern is chromium-6 contamination, which affects the water supply of up to seventy million people around the country. After investigating water pollution around Hinkley, California, *NewsHour* science correspondent Miles O'Brien stated, "In the 1950s and sixties, Pacific Gas & Electric Company admits it dumped twenty-six tons of a coolant made of chromium-6 into unlined retaining ponds here. The poisoning of water in Hinkley was portrayed in the film *Erin Brockovich*."

The chemical is toxic and causes cancer. PG&E has spent $700 million to clean up the water supply around Hinkley, but while the chemical's presence has been reduced, it still remains in the groundwater. With mounting evidence that chromium-6 is more dangerous than once thought, the EPA has decided to revisit their standards for the amount of the chemical that is allowed into the drinking water supply. In April 2014, California's Public Health Department adopted the nation's first drinking water standard for chromium-6 (hexavalent chromium). The standard was set at ten parts per billion, the equivalent of about ten drops of water in an Olympic-sized swimming pool. This is five hundred times greater than the level set be the state's Environmental Protection Agency (EPA).

Unfortunately, after years of contamination, it may already be too late. Such hazardous water threatens the health of communities all across the country. In Massachusetts alone, more than a third of the towns have lost all or portions of their drinking water to toxic contamination, according to the Toxic Action Center (TAC), a New England–based group dedicated to helping communities cope with hazardous pollution resulting from more than a century of irresponsible and illegal handling of toxic chemicals.

According to the TAC, the most common public health threat is hazardous waste contaminating drinking water. "Toxins can seep into buildings built near hazardous waste sites, causing indoor air problems, respiratory diseases, and chemical sensitivity," stated the TAC website, adding, "The experience of living in a contaminated home not only ends normal life, but also can cause serious psychological illnesses."

And it's not only industrial waste that arouses concern. We've previously seen how toxic many prescription medications truly are. And every day, huge amounts of these drugs are flushed away into the water system. Just a partial list of drugs found in water supplies is highly alarming. It includes antidepressants, anticonvulsants, tranquilizers, antibacterials, antipsychotics, ACE inhibitors, nitroglycerin, and steroids.

According to a 2010 investigation by *Time* writer Jeffrey Kluger, there are about three thousand prescription pharmaceuticals being used in the U.S. and thousands more over-the-counter drugs, "not to mention creams and ointments we smear on and then shower off." John Spatz, commissioner of Chicago's department of water management, observed, "Between cosmetics, pharmaceuticals, and other sources, there are eighty-thousand potential combinations of chemicals."

Kluger noted that while wastewater from homes gets treated at sewage plants, it's never possible to remove every trace of drugs. "What's more, sewage pipes break, septic tanks overflow, and in some parts of the U.S. 'straight-piping'—which sends untreated sewage flowing directly into surface water—is still practiced. One way or another, the drugs find their way back to us," he wrote.

Even everyday items, such as the new "energy-saving" mercury-filled coil lightbulbs, may cause health problems. Compact fluorescent lights (CFLs) are said to be safe, but only if the glass is tube is not broken. Cracked or shattered bulbs can release as many as four to five milligrams of mercury, enough to contaminate six thousand gallons of water.

Clearly, clean water needs to become a national priority. The Toxic Action Center suggests a federal policy that would identify hazardous waste sites and levy a pollution tax against any company dumping toxic materials. It also stresses that states should provide technical assistance, adequate funding, and aggressive deadlines to affected residents in order to ensure the purity of their water.

Without this sort of action, we are already beginning to see the dire consequences. In August 2014, five hundred thousand citizens of Toledo, Ohio, were left without water due to reports that contaminated algae had been found in Lake Erie, the city's primary water supply. To complicate this situation, officials said the common practice of boiling water for safety only made the toxic algae more concentrated.

The corporate mass media explained the water contamination as the result of "algae blooms" in Lake Erie but mostly failed to mention the true cause, which, according to the Lake Erie Ecosystem Priority (LEEP), a creation of the water conservation group International Joint Commission (IJC), was phosphorus and other chemicals from both agricultural and urban sources.

An obvious solution to such contamination would be to end the use of agricultural chemicals that end up in the water supply. But this is rarely mentioned by a media dependent on chemical and pharmaceutical advertising. "It's frustrating that all attempts to stop the poisoning of our planet are characterized as 'leftist agendas' by conservative business publications," wrote Mike Adams. "In my view, political affiliations don't matter if we're all dying from the collapse of a global ecosystem that we destroyed with our own foolish ignorance . . . If I had to pick a philosophical belief system that I really feel strongly about, it would be a system that declares all life to be sacred and seeks to protect living ecosystems from the poisoners who are systematically destroying it."

And in many parts of the country, what little clean water remains is fast disappearing because of unregulated usage. In August 2014, the residents of the San Joaquin Valley in drought-plagued California found themselves without water as individual wells dried up. The

Tulare County Office of Emergency Services resorted to shipping a twelve-gallon-per-person water ration to hundreds of homes without water. Bottled drinking water, enough for about three weeks, was delivered by firemen, the Red Cross, and volunteer groups at a cost of $30,000 and was seen as merely a temporary fix. Some residents told the local media they were reluctant to admit to being waterless out of fear their landlords would evict them or social workers would take their children away. Farmworker Oliva Sanchez said she still gets a trickle from her tap, but dirt started coming out with the water. "I try to use the least possible. I'll move if I have to," she said.

The situation in California is neither localized nor temporary. One state-owned well near Sacramento recorded a one-hundred-foot drop in the water table over three months in 2014, while many other wells simply dried up.

It was no joke on April 1, 2015, when California governor Jerry Brown, citing the ongoing water crisis, ordered mandatory water use reductions for the first time in the state's history. Brown ordered a 25 percent water reduction on the state's four hundred local water supply agencies, which prompted some to recall that during a 1977 drought, a "model" ordinance was proposed that would penalize repeat water wasters with fines up to $300 or thirty days in jail or both.

Richard Howitt, one of the authors of a University of California at Davis study of the water problem, described the state's water woes as a "slow-moving train wreck." He added, "A well-managed basin is used like a reserve bank account. We're acting like the super rich who have so much money they don't need to balance their checkbook."

WATER FOR THE FEW

IT'S CLEAR THAT WATER IS BECOMING INCREASINGLY UNAVAILABLE TO ordinary people, while the world's water supply is rapidly coming under the control of multinational banks and multibillionaires. "Water is the oil of the twenty-first century," declared Andrew Liveris, the chief executive of Dow, a chemical company.

Megabanks such as Goldman Sachs, JPMorgan Chase, Citigroup, UBS, Deutsche Bank, Credit Suisse, Macquarie Bank, Barclays Bank, the Blackstone Group, Allianz, and HSBC Bank are reportedly consolidating control over water resources, even transcending national boundaries to partner with each other to buy up not only water rights and water treatment technologies, but also to privatize public water utilities and infrastructure. At the same time, governments are being pressured to pass laws limiting citizens' rights to water and self-sufficiency.

Oregon resident Gary Harrington was arrested for collecting rainwater and snow on his rural property. Authorities accused him of constructing three "illegal reservoirs" on his 170-acre property. Harrington said although the reservoirs are stocked with largemouth bass, they serve as a contingent against wildfires. "It's totally committed to fire suppression," he explained to the media.

Initially the state allowed Harrington to collect water but reversed its decision in 2003, citing a 1925 law stating the nearby city of Medford has rights to Big Butte Creek and its tributaries. Harrington argued that his water came only from rainfall and snowmelt. The disagreement evolved into a protracted court battle over property rights and government bullying. "When something is wrong, you just, as an American citizen, you have to put your foot down and say, 'This is wrong; you just can't take away any more of my rights and from here on in, I'm going to fight it,'" explained Harrington. Nonetheless, he was found guilty. He surrendered himself in July 2014 to begin serving a thirty-day jail sentence and was also fined $1,500.

Other states are following Oregon's lead with harsh crackdowns on citizens. In July 2014, the California State Water Resources Control Board passed "stage one" emergency regulations that empowered all local water agencies to levy a fine of $500 per violation to anyone caught using more than the allocated amount of water. In Santa Monica, the city council was considering an ordinance that would allot sixty-eight gallons per person for a four-member family, while other locations were considering as few as fifty gallons per person. Water usage would be monitored by satellites as well as meters.

Ecological engineer Jo-Shing Yang, author of *Solving Global Water Crises: New Paradigms in Wastewater and Water Treatment*, said the real story in the global water issue is one involving "interlocking globalized capital." "Wall Street and global investment firms, banks, and other elite private-equity firms—often transcending national boundaries to partner with each other, with banks and hedge funds, with technology corporations and insurance giants, with regional public-sector pension funds, and with sovereign wealth funds—are moving rapidly into the water sector to buy up not only water rights and water-treatment technologies, but also to privatize public water utilities and infrastructure."

The corporate world has gained control over water in a number of ways. One example is the 2007 acquisition of UK water utility Southern Water by JPMorgan Chase. Of note, the oldest living member of the Rockefeller family, David Rockefeller—who has served as CEO of the Chase Manhattan Corporation and joined Chase Bank in 1946, long before it became JPMorgan Chase—is a member of the elite and secretive Bilderberg Group, Council on Foreign Relations, and The Trilateral Commission. In January 2012, the China Investment Corporation bought a 8.68 percent investment in Thames Water, the largest water utility in England, serving parts of the Greater London area, Thames Valley, and Surrey. In November of that year, the Abu Dhabi Investment Authority (ADIA), one of the world's largest sovereign wealth funds, purchased 9.9 percent of Thames Water.

Billionaires also are buying up water supplies. Corporate raider T. Boone Pickens reportedly owns much of the Ogallala Aquifer, the midwestern water source that provides freshwater for the production of roughly one-fifth of the wheat, corn, cattle, and cotton in the United States. Warren Buffett holds nine million shares of the Nalco Holding Company, which was named 2012 Water Technology Company of the Year. This subsidiary of Ecolab makes water treatment chemicals and membranes. "But the company Nalco is not just a membrane manufacturer; it also produced the infamous toxic chemical dispersant Corexit which was used to disperse crude oil in the aftermath of BP's oil spill in the Gulf of Mexico in 2010. Before being sold to Ecolab, Nalco's parent company was Blackstone Group [cofounded by Peter G. Peterson, chairman emeritus of the Council on Foreign Relations]," notes Yang.

In 2005 and 2006, it was briefly reported in the mainstream media that the Bush family had purchased 298,840 acres of land in Paraguay. Not widely reported was that the Bush family land sits over the Guarani Aquifer, a freshwater source larger than Texas and California combined. The Guarani is considered the largest single body of groundwater in the world.

"Unfortunately, the global water and infrastructure-privatization fever is unstoppable," explained Yang, who noted that many local and state governments are suffering from revenue shortfalls and are under financial and budgetary constraints. They will find it hard to refuse the offers of private money from elite banks.

"The elite multinational and Wall Street banks and investment banks have been preparing and waiting for this golden moment for years," said Yang. "Over the past few years, they have amassed war chests of infrastructure funds to privatize water, municipal services, and utilities all over the world. It will be extremely difficult to reverse this privatization trend in water."

Such ownership of water supplies gives the wealthy elite and big banks not only huge profits but worrisome control over this vital liquid.

And despite the fact that water is in such short supply in so many places, many frivolous sources still consume huge amounts of this precious resource. For example, there are 125 golf courses in Palm Springs, California, alone. Despite efforts by the U.S. Golf Association (USGA) to mandate the use of reclaimed water and grasses that don't require as much watering, these courses, located in a desert environment, each use an estimated million gallons of water a day to maintain the fairways and greens. This is about the same amount of water an American family of four uses in four years.

With the quality and availability of water raising concern among both scientists and average citizens, conservation and eliminating the adulteration of our water supply by wastes must become a national priority.

Water issues are a top priority for lawmakers in the states where water shortages are approaching a crisis, but most voters seem oblivious as long as water still flows from their faucets. At least this was the finding of a poll in drought-stricken Texas. According to a University of Texas/*Texas Tribune* poll taken in 2013, just 6 percent of respondents said water should be the top priority, while 50 percent said it should be one of the key priorities and 23 percent deemed it merely a secondary priority.

"The Legislature has water on the brain, so to speak, but it doesn't seem that the public is following," said Jim Henson, codirector of the poll. "If you look at the most important problems facing Texas, water still doesn't move the needle."

FLUORIDE

THOUGH COMMERCIAL WASTE, CHEMICALS, AND DRUGS IN OUR water supply are significant concerns, thus far the adverse reactions to these contaminants have been relatively limited in scope. But another, far more ubiquitous concern is impossible to ignore. Today, more than 70 percent of Americans are drinking water containing sodium fluoride, a toxic byproduct of the aluminum industry.

Despite the continued protestations of the American Dental Association and various U.S. government agencies, science has clearly demonstrated that fluoride is a toxic chemical that can accumulate in the body and brain, causing harm to enzymes and producing serious health problems, including neurological and endocrine dysfunction. Children are particularly at risk for adverse effects of overexposure to fluoride.

The fluoridation of drinking water began in Grand Rapids, Michigan, in 1945 at the instigation of Drs. H. Trendley Dean and Gerald J. Cox, both employees of Andrew W. Mellon, founder of the Aluminum Company of America (Alcoa), who as U.S. treasury secretary in 1930 oversaw administration of the Public Health Service (PHS). Mellon had Dean, who worked for the PHS, study the effects of naturally fluoridated water on teeth. Dean reported that while fluoride caused discoloration of teeth, in some cases it might prevent cavities. Meanwhile, as a researcher at the Mellon Institute in Pittsburgh, Cox was assigned to study the effect of fluoride on the teeth of lab rats. He concluded that it seemed to slow decay, and in 1939, he proposed to fluoridate the entire public water supply.

Sodium fluoride is produced by adding fluorine, a highly toxic pale yellow diatomic chemical. It's worth considering the other uses for fluorine. It can be found in compounds called organofluorines, carbon-fluorine-bonded chemical compounds used to produce such items as Teflon, Gore-Tex, and many drugs such as Prozac (fluoxetine), Cipro (ciprofloxacin), and Baycol (cerivastatin). Sodium fluoride also has been used as a rat and bug poison, fungicide, and wood preservative. Not only is fluoride a product of the aluminum industry, it has also been used in the manufacture of atomic bombs. This waste accumulated after World War II in the wake of atomic bomb testing. Critics of fluoride have been called conspiracy theorists and shills for junk science, but have gained credibility in light of recent peer-reviewed studies.

Dr. Donald Miller, a cardiac surgeon and professor of surgery at the University of Washington in Seattle, explained how this poison was made palatable to the public. "With several instances already on record of fluoride causing damage to crops, livestock, and people downwind from industrial plants, government and industry, led by officials running the Manhattan Project, sought to put a new, friendlier face on fluoride. This would dampen public concerns over fluoride emissions and help forestall potentially crippling litigation. Instead of being seen as the poison it is, people should view fluoride as a nutrient, which gives smiling children shiny teeth, as epitomized in the jingle that calls fluoride 'nature's way to prevent tooth decay.'

"It worked. Early epidemiological studies showed a 50 to 70 percent reduction in dental cavities in children who drank fluoridated water. These studies, however, were poorly designed. None were blinded, so dentists examining children for caries [cavities] would know which kind of water they were drinking. Data gathering methods were shoddy. By today's evidence-based medicine standards these studies do not provide reliable evidence that fluoride does indeed prevent cavities," wrote Dr. Miller. In other words, we're being forced to ingest a substance that can damage the brain, lower IQ, and have neurotoxic effects, and we're not deriving any benefit from it.

Pro-fluoridation scientists, along with dental professionals and public health officials, continue to insist that fluoridation of water, toothpaste, and mouthwash is generally safe and can significantly reduce cavities and tooth decay. According to the American Dental Association (ADA), water fluoridation reduces tooth decay by 20 to 40 percent with the only significant negative consequence being the risk of dental fluorosis, a discoloration of tooth enamel that occurs with higher rates of exposure to fluoride. They say this is only a cosmetic issue, not a health risk. But those who have investigated fluoride closely know otherwise.

A study by chemists from Russia and Australia published in the *Journal of Analytical Chemistry* in May 2014 indicated that fluoride ions found in fluoridated water and toothpaste may lead to an increase in urinary stone disease (USD). Researchers studied twenty urinary stones from Russian hospital patients and discovered fluoride ions in 80 percent of them. They concluded this could be traceable to high levels of fluoride in the patients' urine, possibly a result of drinking water containing fluorides and ingesting fluoride toothpaste. Other studies found on the Internet also link excessive amounts of fluoride to kidney stones (nephrolithiasis).

Based on such recent studies, the antifluoride movement has been gaining strength among informed citizens. In 2014, the Dallas City Council said it would consider removing fluoride from the city's water supply after hearing arguments from fluoride opponents. Regina Imburgia, the founder of Activists for Truth in Dallas, says the council had heard statements regarding fluoride hazards several times in the past but "not one member responded." She says she was gratified that there now was some movement on "this very serious issue."

As it always does, the mainstream media responded to the antifluoride initiative with sarcastic disdain. "Anti-Fluoride Cranks at City Hall: Is it Something in the Water?" chided one newspaper headline, while the Dallas County Dental Society issued the following statement: "We believe the claims and tactics used by fluoride opponents are not founded in research but fear." Such derision produced results, for in early 2015 the city council voted to continue the fluoridation of the city's water.

Yet fluoride critics point to a wide body of research supporting their claims, including a 2010 book by Paul Connett, James Beck, and Spedding Micklem entitled *The Case Against Fluoride: How Hazardous Waste Ended Up in Our Drinking Water and the Bad Science and Powerful Politics That Keep It There*. Supported by eighty pages of references, the book underscores the author's conclusions that the fluoride in our drinking water is the hazardous byproduct of the phosphate, fertilizer, and aluminum industries. It is illegal to dump this waste into the sea or local surface water, and yet it is allowed in our drinking water. Nonmedical city workers dump fluoride into the water supply without regard for dosage or quality. Even more damning, the addition of fluoride constitutes medication without informed consent.

"Whatever the reasons that led the U.S. PHS [Public Health Service] to endorse fluoridation in 1950, researchers did not have solid evidence to demonstrate either the short-term or the long-term safety of this practice," concluded the book's authors. "Not only was safety not demonstrated in anything approaching a comprehensive and scientific study, but also a large number of studies implicating fluoride's impact on both the bones and the thyroid gland were ignored or downplayed." Connett addressed the Dallas City Council in June 2014 to outline the various reasons fluoridation is unethical, including the fact that human beings do not need fluoride, and the fact that Americans have no choice but to consume fluoride.

"The primary strategy that promoters of water fluoridation use is to establish that authorities say: 'It's good; it's safe and effective' and that 'people who oppose fluoridation are stupid, stupid, stupid.' This strategy is basically aimed at keeping you from looking at the actual science," said Connett, adding this advice: "The evidence that fluoride causes harm is growing. But before [city] councillors get bogged down in trying to understand which side of the argument has the best or most accurate information, they need to answer the question . . . do they have the right to do what a doctor is not permitted to do—i.e., to medicate people without their consent? No, they do not have the right to force medication. To put a medicine in the drinking water defies many aspects of medicine. You can't control the dose. You can't control who gets it. It goes to everybody, including bottle-fed babies."

According to a 2012 meta-analysis conducted jointly by the Harvard School of Public Health and China Medical University in Shenyang, studies strongly indicated that fluoride may adversely affect cognitive development in children. All but one of the twenty-seven studies, involving more than eight thousand Chinese school-age children, indicated that high fluoride content in drinking water negatively affected cognitive development. Children in high-fluoride areas had significantly lower IQ scores than those who lived in low-fluoride areas.

At least forty-nine studies investigating the relationship between fluoride and human intelligence conducted by November 2014 along with a total of thirty-two studies investigating the relationship of fluoride with animal learning and memory in animals were listed on the FluorideAlert.org website. Out of the forty-nine human studies based on the IQ examinations of more than eleven thousand children, forty-two found reduced IQ associated with elevated fluoride exposure. Thirty of the thirty-two animal studies found fluoride impaired both memory and learning. The studies provided evidence that cumulative exposure to fluoride can cause damage to the developing brain of both children and animals.

Other, more limited work in the U.S., such as a 2006 National Academy of Science report, also concludes that fluorides have the ability to interfere with the functions of the brain and the body by both direct and indirect means. "Fluorides also increase the production of free radicals in the brain through several different biological pathways. These changes have a bearing on the possibility that fluorides act to increase the risk of developing Alzheimer's disease," states the report.

The National Academy of Science report includes recommendations that the EPA lower its limit for fluoride in drinking water from 4 milligrams to 0.7. The report warns that even at 2 milligrams severe fluorosis can occur, and that a lifetime of drinking water with fluoride at the 4-milligram rate could increase the risk of broken bones. Another study in 2006, this one carried out by the National Institutes of Health (NIH), suggests that drinking water with more than two parts per million of fluoride can cause damage to liver and kidney functions in children.

The effects on the brain may also be dire. "Fluoride seems to fit in with lead, mercury, and other poisons that cause chemical brain drain. The effect of each toxicant may seem small, but the combined damage on a population scale can be serious," says Philippe Grand-jean, an author of the NIH study.

The studies touting the safety of fluoride rest on shaky ground. After looking at a number of water fluoridation studies, Britain's University of York Centre for Reviews and Dissemination was "unable to discover any reliable good-quality evidence in the fluoridation literature worldwide." The review did not show water fluoridation to be safe. In fact, it found the quality of the research was too poor to establish with any confidence the safety of fluoridation.

Connett and his coauthors note that the dental industry is not entirely responsible for the inadequate information on fluoridation. "The enemy, instead, is a system that is geared to perpetuate a belief in fluoridation by using professional peer pressure, tremendous amounts of promotional money, and the subtle threat of ostracism. Maintenance of the policy of fluoridation continues by use of the tools of professional licensure and legal pressure in a long chain of workers who are compelled to continue a practice many of them know is wrong. And, some are compelled to follow state mandates that have been poorly understood, poorly written, yet strongly enforced by professionals who were responsible for twisting laws and ethics that should have signaled a poor and failing medical policy many years ago," they wrote. "Only the politics of face-saving seems to have formed the basis for the continuation of fluoridation, especially in the U.S. The ignorance by officials of the basic science, law, and ethics concerning this issue are unbelievably glaring."

And the problems already described pale in comparison to the potential effects fluoride may have on the thyroid gland, which is one of the body's largest endocrine glands and controls how quickly the body uses energy, makes proteins, and processes hormones. A 1958 article in the *Journal of Clinical Endocrinology* entitled "Effect of Fluorine on Thyroidal Iodine Metabolism in Hyperthyroidism" noted that "thyroidal, blood and urinary radioiodine studies suggest that fluorine inhibits the thyroid iodide-concentrating mechanism." More recent studies, including a 2006 National Research Council (NRC) report entitled "Fluoride in Drinking Water: A Scientific Review of EPA's Standards," concluded that further research is required on the effects of fluoride on endocrine function, especially with respect to a possible role in the development of several diseases or mental states.

According to the NRC fluoride report, which was produced by a panel of experts who reviewed hundreds of published fluoride studies, even small amounts of fluoride consumed from tap water not only can disrupt thyroid function but can damage bones, teeth, and brain, and even lower IQ and/or cause cancer.

Further studies may offer an explanation for many who suffer from hyperthyroidism, a condition in which the thyroid gland produces too much of the hormone thyroxin, causing such problems as irregular heartbeat (arrhythmia), heart palpitations, nervousness, anxiety and irritability, tremors, sweating, increased sensitivity to heat, goiters, sleeping difficulties, thinning of the skin, fatigue, and muscle weakness. It also begs the question of the damage fluoride may be capable of inflicting on a normal thyroid.

The team of Dr. Richard L. Shames and his registered-nurse wife, Karilee, in 2013 changed their attitude regarding fluoridation and the thyroid after noticing the unexplained skyrocketing of thyroid disease and its spin-off epidemics of fatigue, depression, anxiety, infertility, and obesity.

The Shameses discovered a lengthy list of reports in medical journals from around the world regarding the harmful effects of fluoride. They also learned that prior to the advent of such thyroid treatment drugs as Tapazole, fluoride was used by the medical profession to slow overactive thyroid conditions. "Every medicine has a good action, called 'the benefit,' and other less desirable actions called 'side effects.' In hindsight, it did seem odd that fluoridated water was the only substance ever discovered that had a great benefit with no side effects at all. Once we thought about it carefully, it also seemed curious that fluoride was the only medicine ever to be added to public drinking waters."

They found, after reviewing hundreds of articles and books, that fluoride could indeed be considered a "hormone disruptor," that class of chemicals from many unrelated sources that have the unintended consequence of altering the proper function of important hormones in the body, such as that produced by the thyroid.

They wondered if perhaps the low concentration of fluoride in water supplies could help prevent tooth decay without harming the thyroid. They found that the data indicated otherwise.

Although the controversy over water fluoridation most likely will continue for years, it is now abundantly clear that the benefits of fluoridation have been overblown, while the risks have been understated. And the substance is ubiquitous: fluoride can be found in a myriad of substances besides toothpaste and water. These include fruit juices, soda pop, tea, and processed foods; even California wines, the grapes of which are sprayed with the pesticide cryolite (sodium aluminum fluoride). In other words, fluorine is everywhere, and consumers must be vigilant in order to avoid consuming it.

CHANGING THE GAME

POTABLE WATER IS A FUNDAMENTAL NECESSITY FOR LIFE AND HEALTH. The public must become more aware of the increasing scarcity of water, the pollution of many water supplies, as well as the control being sought by large multinational corporations such as Coca-Cola, Perrier, and Nestlé along with major financial institutions. Frivolous use of water for spacious lawns and golf courses, especially in arid climates, must be reevaluated in light of the growing water shortage.

The crisis over clean water and, in fact, the availability of water itself is a recent phenomenon. Years ago, if you asked for water in a grocery store, you would be pointed to the drinking fountain. Today entire aisles are filled with bottled water, which considered by the gallon costs more than gasoline. The majority of these bottles are plastic and, along with aluminum can lining, contains the endocrine disruptor Bisphenol-A (BPA), itself the object of much controversy concerning potential health hazards.

With the giant corporations that hold sway over government regulators profiting from the sale of water and water-related enterprises, it is little wonder that neither they nor the federal government has given clean water a top priority. Such lethargy on the issue of water is allowed by a population born into a society that accepts water from the tap as an unquestioned right.

The public must be more mindful of water pollution. No longer can the public afford to continue discarding insoluble materials, to include medical waste, cosmetics, and household cleaners, into sinks and toilets.

The practice of introducing fluoride into the nation's water supplies must cease, if for no other reason than it is improper to medicate an entire population without their consent. Additionally, recent studies are revealing a number of problems associated with long-term consumption of fluoride, including thyroid damage and a susceptibility to cancer.

Some have come to view the ubiquitous fluoridation of water, toothpaste, and other consumables as a covert attempt by the globalist elite to both dumb down and reduce the human population.

For those who feel a need to use fluoridated products for their children's teeth, as with many health matters, moderation is the key. Use it sparingly and under close supervision, and do not allow children to swallow products containing fluoride.

CHAPTER 14

DEADLY AIR

IN THE UNITED STATES TODAY, EVEN THE VERY AIR WE BREATHE can be hazardous.

In 2012 alone, more than seven million people worldwide died as a result of air pollution. Some of this pollution came from likely sources—ill-regulated industries and carbon-spewing vehicles—but it also came from less obvious sources like stoves and cooking fires. Statistics from the World Health Organization, along with a World Bank study issued in association with China's Development Research Center, identify air pollution as the world's single biggest environmental health risk.

Both reports observe that the burning of coal, wood, and animal waste for fuel is one of the greatest threats to human health. And it is estimated that with the increasing urbanization of the human population, especially in China, high levels of air pollution will continue to grow. Alarmed Chinese officials in early 2014 declared a "war on pollution" after Beijing continually recorded "very unhealthy" air quality. The pollution got so bad that the British School of Beijing installed a giant inflatable dome, complete with air locks, for sports activities.

Americans too are affected by unrestrained air pollution. Research by the National Academy of Sciences (NAS) published in

2013 notes, "Outsourcing production to China does not always relieve consumers in the United States—or for that matter many countries in the Northern Hemisphere—from the environmental impacts of air pollution."

Coauthor Steven J. Davis said study researchers detected a decline of air quality in the western U.S. caused by dangerous spikes in contaminants being carried across the Pacific by wind currents. They noted dust, ozone, and carbon accumulated in valleys and basins in California and other western states. "We've outsourced our manufacturing and much of our pollution, but some of it is blowing back across the Pacific to haunt us," remarked Davis.

Any product, no matter where it was created, can cause pollution. Each nation must assume some responsibility for harmful emissions whether the end product is produced there or elsewhere.

By mid-2014, even officials of the federal government acknowledged the danger posed by air pollution. In June of that year, EPA administrator Gina McCarthy announced new rules intended to reduce air pollution by requiring existing coal-fired power plants to reduce carbon-dioxide emissions by 30 percent by 2030. She said, "Carbon pollution from power plants comes packaged with dangerous pollutants, like particular matter, nitrogen oxide, and sulfur dioxide, and they put our children and our families at even more risk." She added, "If your kid doesn't use an inhaler, you should consider yourself a very lucky parent, because one in ten kids in the U.S. suffers from asthma." McCarthy said the new EPA rules on primarily coal-fired power plants would be an "investment in better health and in a better future for our kids," and would actually lower electric bills.

However, critics claimed reducing carbon-dioxide pollution in the U.S. would not impact climate-changing pollution from Asia and other developing areas.

Despite legislation such as the Clean Air Act, which has helped reduce air pollution, fossil-fuel power plants, boilers, and cement plants still belch pollutants into the sky: the total public cost of pollution in the U.S. was estimated as recently as 2007 at more than $200 billion yearly. Environmental organizations such as the Natural Resources Defense Council argue that until stronger standards to reduce toxic emissions from coal and fossil-fuel-burning industries are implemented, harmful toxic chemicals will continue to be released into the air of our communities, threatening public health.

RADIATION

IF THE POLLUTED AIR ALONG WITH DEADLY FOOD AND WATER DON'T kill us, man-made radiation just might. As if the destructive power of oil isn't enough, humanity now has to contend with atomic weapons and leaking radiation from nuclear power plants.

This issue was brought to the forefront following the devastation of Japan and the Pacific due to the ongoing leakage of deadly radiation from the 2011 tsunami-damaged Fukushima Daiichi nuclear reactors. When the reactors were damaged by the 2011 tsunami, the world was brought to a near-extinction-level event and the facility continues to spew dangerous radiation although one has not heard much of this recently in the corporate mass media.

The Fukushima number one plant, located about 136 miles northeast of Tokyo, suffered three reactor core meltdowns after power to cooling pumps was lost following the March 2011 earthquake and tsunami. It was the world's worst nuclear disaster since the explosion and fire at the Russian Chernobyl plant released radiation into the air in 1986.

Since the 2011 tsunami, an ongoing series of misrepresentations and blunders have plagued the crippled plant. Much reliance was placed on a state-of-the-art advanced liquid processing system (ALPS) project, designed to remove the most dangerous nuclides. However, the ALPS system proved functional only during periodic tests.

In September 2013, the Japanese government and Tokyo Electric Power Company (TEPCO) officials admitted that radiation readings around the power plant were eighteen times higher than they originally reported. In February 2015, Fukushima plant sensors detected a new leak of highly radioactive water being dumped into the sea. This water contained radiation contamination levels up to 70 times greater than previous leaks. By early 2014, radiation around the Fukushima plant had reached eight times government safety guidelines, according to TEPCO, and the power company was still struggling to contain radioactive leaks. TEPCO attempted to reduce the radiation level by injecting water into the three crippled reactors, but this only created highly contaminated water. TEPCO's response has been plagued with problems. More than one hundred tons of radioactive water spilled from a container during operations in early 2014. Officials said the irradiated water was erroneously thought to be contained within a building.

Akira Ono, manager of the Fukushima number one nuclear power plant, later admitted that repeated efforts to control the radioactive water had failed. The radioactive water was moved to the wrong building. Ono's admission came eight months after Japanese prime minister Shinzô Abe announced the problem had been resolved.

Previously, the Japanese government had begun diverting some of the four hundred tons of groundwater pouring into the complex each day into the ocean after gaining the approval of fishermen. Even then, elevated levels of radiation were found in the groundwater, although TEPCO officials said it was within permissible limits.

A report presented at a conference of the American Geophysical Union's Ocean Sciences Section in February 2014 indicated that some Caesium-134 had already arrived in Canada's Gulf of Alaska area. Radiation carried by both air and sea was expected to reach the west coast of the U.S. by mid-2014, but authorities said it would only be at very low levels. But some were questioning if any rise in radiation levels might prove harmful.

Some Japanese university researchers have reported their schools will not give them funds or support for research involving the Fukushima facility. Biologist Joji Otaki with Japan's Ryukyus University, who has authored papers on deformities in butterflies caused by Fukushima radiation, advised, "Getting involved in this sort of research is dangerous politically."

Michio Aoyama, a senior scientist with the Japanese government's Meteorological Research Institute, reported radioactive Caesium-137 in the surface water of the Pacific Ocean estimated at levels ten thousand times higher than nuclear contamination measurements from the Chernobyl nuclear accident.

Aoyama submitted an article reporting these alarming radiation levels to the publication *Nature*. But when he complied with the request of his institute's director general to remove his own name from the article, the piece was canceled.

Such censorship has not been limited to Japan. Timothy A. Mousseau, a professor of biological sciences at the University of South Carolina, attempted to research Fukushima radiation but said he was hampered by the Japanese government and three of his associates in the testing dropped out over concerns their future job prospects might be jeopardized.

The fear that truthful information on radiation was being kept from the public was heightened by the fact that no federal agency was monitoring Pacific Coast seawater for radiation. "I'm not trying to be alarmist. We can make predictions, we can do models. But unless you have results, how will we know it's safe?" asked Ken Buesseler, a chemical oceanographer at Oregon's Woods Hole Oceanographic Institution.

No one is suggesting the 2011 earthquake and tsunami were the result of some conspiracy, but questions remain over the lack of immediate and effective response, not to mention the ongoing lack of serious coverage in the mainstream corporate mass media. Is this just laziness and inattention on the part of the corporate media or is it indicative of the globalists who want to maintain the status quo while culling the population?

Adding to radiation concerns is the fact that some insurance companies in 2014 began notifying customers that their company would no longer cover any claims relating to nuclear energy. Traveler's is reported to have announced a new nuclear-energy-liability-exclusion endorsement that includes "all forms of radioactive contamination of property."

The change of insurance policy prompted many to ask if the nuclear disaster at Fukushima is as harmless to Americans as the government and some scientists are telling us, then why are insurance companies specifically excluding coverage for nuclear-energy-related claims?

One answer to this question is that in fact there is no safe level of radiation exposure, whether through contact, ingestion, or breathing. Dr. Jeff Patterson, a past president of Physicians for Social Responsibility, stated, "There is no safe level of radionuclide exposure, whether from food, water, or other sources. Period."

Fears are also rising that Fukushima radiation is killing life in the Pacific Ocean. In 2013, Australian yachtsman Ivan Macfadyen sailed the same course from Melbourne to Osaka, Japan, as he had ten years previously. His report to the *Newcastle Herald* was both eerie and frightening.

"I've done a lot of miles on the ocean in my life and I'm used to seeing turtles, dolphins, sharks and big flurries of feeding birds. But this time, for three thousand nautical miles there was nothing alive to be seen," explained Macfadyen. "In years gone by, I'd gotten used to all the birds and their noises. They'd be following the boat, sometimes resting on the mast before taking off again. You'd see flocks of them wheeling over the surface of the sea in the distance, feeding on pilchards." But on this trip, there were no birds, no fish, hardly any signs of life.

Macfadyen says the next leg of his trip from Osaka to San Francisco was even worse. It was a voyage tinged with nauseating horror and a degree of fear. "After we left Japan, it felt as if the ocean itself was dead," he recalled. "We hardly saw any living things. We saw one whale, sort of rolling helplessly on the surface with what looked like a big tumor on its head. It was pretty sickening."

Macfadyen and his brother Glenn, who joined him for the run from Hawaii to the U.S., said not only did they suspect hazards from radiation but they encountered an unbelievable amount of garbage— "thousands on thousands" of yellow plastic buoys, huge tangles of synthetic rope, fishing lines and nets, and pieces of polystyrene foam by the millions. Also slicks of oil and gas. "The ocean is broken," said Macfayden upon returning home.

In 2014, Dana Durnford and Terry Daniels, making a 125-mile voyage up the British Columbia coast from Vancouver, found tidal pools devoid of life. "All the species in the tidal pools are missing . . . We liked to take pictures of the varied life but this time we couldn't find anything to take pictures of," reported Durnford in a radio interview. "I was also looking for bees but I couldn't find but one."

Dr. Timothy A. Mousseau, the aforementioned professor of biological sciences at the University of South Carolina, in August 2014 supported the observations of Durnford and Macfadyen by telling the Foreign Correspondents' Club of Japan, "The [Fukushima radiation] effects on species richness or biodiversity are even more striking, dropping off with increasing radiation . . . Very, very striking patterns of results . . . I would suggest that what it means is that, contrary to governmental reports, there's now an abundance of information demonstrating consequences—in other words, injury—to individuals, populations, species, and ecosystem functions, stemming from the low-dose radiation due to Chernobyl and Fukushima disasters."

When asked by a reporter if the reported disappearance of animal life might be an effect of radiation on the entire ecosystem, Mousseau replied, "Yes, I think the only conclusion you can come to from the increasing body of evidence of Chernobyl [and Fukushima] is that all components of this ecosystem seem to be affected, from the bacteria in the soil, the fungi in the soil, all the way up to the top predators . . . they are all connected of course. As we pick away at the various components of the ecosystem, we have not found any particular components that don't seem to be affected in some way."

It would appear that the corporate giants put more effort into the construction and financing of nuclear power plants than in determining suitable locations that would provide stable understructure and the least impact on the environment.

Animal and plant life were not the only things to suffer ill effects in the wake of the Fukushima disaster. The U.S. aircraft carrier *Ronald Reagan* was sent to provide aid to the Japanese. As of 2014, several dozen crew members were still embroiled in a court dispute over damages for health problems they claim were caused by exposure to excessive radiation.

Still, so far Japan has borne the brunt of the radiation. One Tokyo physician, Dr. Shigera Mita, countered government and

TEPCO assurances that Fukushima was under control by leaving the city in March 2014, warning that everyone in eastern Japan has become a victim of radiation. He said his clinic has carried out blood examination and thyroid ultrasound examination on 1,500 patients, many of them children.

In a letter to fellow Japanese doctors, Dr. Mita wrote that his clinic had witnessed increased cases of sinusitis. He said many such cases were accompanied by asthma lasting longer than in the past. He said when children spent at least two weeks in western Japan, they recovered. "With elderly people, it takes more time for asthma to heal. The medication doesn't seem to work." Dr. Mita said his clinic was seeing more patients with diseases that had been rare before, such as polymyalgia rheumatic, a disease common among persons over age fifty and formerly contracted by 1.7 people out of every 100,000. "Before [March 2011], we had one or less patient per year. Now we treat more than ten patients at the same time," he said, warning, "Ever since [March 2011], everybody living in eastern Japan including Tokyo is a victim, and everybody is involved . . . The key word here is 'long-term low-level internal irradiation.'"

There has been a discernible lack of radiation studies today by the federal government perhaps because leaders are disinclined to remind the public of the more than four thousand secret radiation experiments on U.S. citizens, including some two hundred thousand "atomic vets" (soldiers exposed to atomic testing). These tests, made public only in recent years, produced "downwinders," residents of Nevada, Utah, Colorado, New Mexico, and West Texas who were exposed to radioactive fallout from the testing of more than two hundred atmospheric and underground tests. Such tests were conducted under the auspices of the Atomic Energy Commission, the Defense Department, the Department of Health, Education and Welfare, the Public Health Service (now the CDC), the National Institutes of Health, the Veterans Administration (VA), the CIA, and NASA.

In the 1950s and sixties, radiation meters, Geiger counters, and gas masks were routinely provided by Civil Defense. Today, they are relics and difficult to acquire.

In view of the lack of governmental preparedness and research along with worrisome reports such as those from Dr. Mita, private citizens have begun taking it upon themselves to monitor for radiation. Christina Consolo, a radiation expert and Canadian radio host known as "Radchick," claimed important information on radiation was being kept from the public by both federal authorities and the airlines.

Consolo said both she and her daughter in 2013 suffered kidney failure after absorbing one-tenth of the FAA-accepted yearly radiation exposure level on a flight from Canada to Cancún, Mexico. She said she had taken a Geiger counter with her on that trip and that the TSA started impounding Geiger counters soon afterward. Consolo said her research showed a rise in cases of pilots passing out at the controls, sicknesses among both crews and passengers following flights, and increased instances of unruly passengers.

Although the TSA website does not list Geiger counters or radiation detectors as prohibited items, a number of personal experiences by air travelers found on the Internet make it clear that this may be up to the prerogative of the individual TSA agent and one should be most careful and discreet when boarding an airliner with such a device.

Reports on radiation in California have been mixed. In 2012, scientists at Stanford's Hopkins Marine Station in Monterey County reported low levels of radioactive Caesium in Pacific Bluefin tuna caught off the coast of San Diego. One San Mateo resident posted a YouTube in early 2014 indicating levels of radiation on a nearby beach at five times the level of normal background radiation. However, state officials were quick to respond. Wendy Hopkins of the California Department of Public Health told the media, "Recent tests show that elevated levels of radiation at Half Moon Bay are due to naturally occurring materials and not radioactivity associated with the Fukushima incident. There is no public health risk at California beaches due to radioactivity related to events at Fukushima." Other health officials joined in, claiming it was safe to visit beaches.

David Hirsch, a nuclear policy expert and lecturer at the University of California at Santa Cruz, quipped, "No one should fear a day on the beach. No one should fear surfing or eating fish."

In mid-May 2014, after analyzing kelp samples from twenty-six locations along the Pacific coast, including samples taken in Hawaii and Guam, a forty-member team headed by Steven Manley, a professor of marine biology at Cal State Long Beach, and Kai Vetter, a professor of nuclear physics and engineering at UC Berkeley, concluded that no signs of Fukushima radiation had yet been found on the West Coast. "According to predictions based on our scientific models, we should see at some point the arrival of small concentrations of Caesium. But the concentration we are expecting is extremely small and most likely won't be a danger to the public," assured Vetter. In other words, the mainstream scientific community is not concerned.

Yet some researchers suggest that certain Washington officials may have known that the Fukushima disaster was worse than initially reported. Furthermore, these skeptics point out, many nuclear reactors in the U.S. are of similar design and could be susceptible to similarly catastrophic damage. Five of the six Fukushima reactors were General Electric (GE) Mark-1s. There are twenty-three GE Mark-1 BWR reactor nuclear plants in the United States.

As far back as 1972, Stephen H. Hanauer, then a safety official with the Atomic Energy Commission, recommended that the Mark-1 system be discontinued because it presented unacceptable safety risks. Hanauer specifically cited the smaller design of the Mark-1 containment vessel (the steel-and-concrete capsule serving as a final protection against cooling loss) as being susceptible to explosion and rupture, just what happened at Fukushima. Joseph Hendrie, who would later become chairman of the Nuclear Regulatory Commission, argued that while a ban on the Mark-1 systems was an attractive possibility, the technology was so widely accepted by the industry and regulatory officials that "reversal of this hallowed policy, particularly at this time, could well be the end of nuclear power." Once again, politics and profits trumped safety because nuclear power generation accounts for nearly 20 percent of U.S. energy needs; costs for plants as currently planned could top $1.6 trillion.

Chris Carrington, writing for Canada's Centre for Research on Globalization, asked why General Electric, which with TEPCO jointly operated the Fukushima facility, was not being held accountable for unsafe U.S. reactors or the Fukushima meltdown. "Here's one possibility," he noted. "Jeffrey Immelt is the head of GE. He is also the head of the United States Economic Advisory Board. He was invited to join the board personally by President Obama in 2009 and took over as head in 2011 when [former Federal Reserve chairman] Paul Volcker stepped down in February 2011, just a month before the earthquake and tsunami that devastated Fukushima.

"There is no way that Immelt doesn't know about all the warning his company was given about the design flaws of the Mark 1; and if he knows, the government knows," said Carrington.

He said Obama must have known that the radiation from Fukushima is worse than it would have been had the reactors used at the plant been of a different design, but considered that "GE cannot afford a corporate lawsuit, and neither can the Obama administration. It wouldn't be pretty if a senior adviser to the president was hauled through the courts. There's a chance it would not just be GE that went down in the wake of such a case."

It should also be noted that a year after the Fukushima meltdown, TEPCO was taken over by the Japanese government, as the company was unable to pay for the damage and repair of the reactors, with this cost estimated as high as $250 billion. Under Japanese law, the companies supporting Fukushima, including suppliers like GE, Hitachi, and Toshiba, are exempt from liability. According to a report by Greenpeace International, nuclear suppliers, including GE, Toshiba, and Hitachi, are involved in the decontamination and decommissioning at Fukushima and are profiting from it. "Governments have created a system that protects the profits of companies while those who suffer from nuclear disasters end up paying the costs," states the Greenpeace website.

The ongoing disaster at Fukushima has had some effect on public opinion, energizing nuclear opponents and even winning over a few formerly pro-nuclear politicians. Naoto Kan, Japan's prime minister during the meltdown, explained how he came to oppose nuclear power while still in office.

After being forced to consider the evacuation of Tokyo, Kan said, "It's impossible to totally prevent any kind of accident or disaster happening at the nuclear power plants. And so the one way to prevent this from happening, to prevent the risk of having to evacuate such huge amounts of people . . . fifty million people, and for the purpose, for the benefit of the lives of our people, and even the economy of Japan . . . I came to change the position, that the only way to do this was to totally get rid of the nuclear power plants."

Kan told radio journalist Amy Goodman his message to President Obama regarding nuclear plants would be: "When considering energy policy from now and considering the issues and the problems of cost and also nuclear waste, while it may have once been said that there was a nuclear renaissance, nuclear technology now is clearly old and dangerous technology, and we need to be looking at other ways." Where nuclear power once was hailed as providing energy too cheap to meter, more and more people are realizing that nuclear power is too expensive (both in costs and risks to humans) to continue supporting.

Though Fukushima is the most notable recent case, the problem of leaking radiation is by no means restricted to Japan.

In March 2006, about nine gallons of highly enriched uranium leaked from a pipe at the privately owned Nuclear Fuel Services (NFS) facility at Erwin, Tennessee. If the leak had continued and puddled, it could have caused a chain reaction and achieved critical mass. Luckily, it was caught and cleaned up before that could happen.

But it took more than a year for the public to be made aware of the leak, and this only occurred after the Nuclear Regulatory Commission (NRC) inadvertently mentioned a uranium leak in its annual report to Congress. The NRC had designated correspondence with the NFS as "official use only," which prevented the incident from being made public, confirming the fact that in many cases the public isn't even aware of the most dangerous radiation threats.

In May 2014, New Mexico Environment Department (NMED) secretary Ryan Flynn revealed a radiation leak in February at the Waste Isolation Pilot Plant near Carlsbad. It was blamed on the mishandling of nuclear waste. Flynn ordered the Los Alamos National Laboratory (LANL), which controls the nuclear waste dump, to submit a plan to secure fifty-seven containers in which nuclear waste was packed with nitrate salts and organic kitty litter for storage. It was believed that exchanging organic kitty litter for less absorptive nonorganic litter may have caused the containment vessel to leak, contaminating twenty-two workers with low-level radiation.

Although officials tried to downplay the danger of the radiation release, Flynn signed a document warning, "Based on the evidence presented to NMED, the current handling, storage, treatment and transportation of the hazardous nitrate salt bearing waste containers at LANL may present an imminent and substantial endangerment to health or the environment." According to the Associated Press, initial investigations into both the container leakage and a recent fire at the plant placed blame on a slow erosion of the safety culture at the fifteen-year-old, multibillion-dollar site. In the dull daily routine of plant operation, workers become complacent and sloppy.

In July 2014, the South Carolina Electric & Gas nuclear power plant northwest of Columbia was shut down for weeks after irradiated water was found leaking from a pressurizer safety valve. In a notice to the Nuclear Regulatory Commission, the utility said the V. C. Summer plant in Fairfield County was closed for repairs after a small amount of radioactive water was found inside it. A utility spokesman said the safety valve was part of the plant's cooling system, keeping the reactor from overheating and causing a radiation release. The plant is next to two new reactors under construction in a $10 billion project.

And even worse than damage from weather or old equipment is the possibility of sabotage, such as occurred at the Doel 4 reactor in Belgium in mid-2014 after an oil leak in a turbine caused the plant to be shut down until at least the end of the year. With an additional two reactors (Doel 3 and Tihange 2) already closed because of cracks in steel reactor casings, Belgium's nuclear capacity was reduced by one half. A spokesman for the plant's owner, GDF Suez, said the damage to Doel 4 was the result of "an intentional manipulation" in that someone had tampered with the system for emptying oil from the turbine. Later, Belgian authorities said the damage may have been an act of terrorism. Very little mention of this possible sabotage was made in the U.S. corporate mass media.

Obviously, the continued use of nuclear power will only prolong the ongoing series of radiation leaks and potential disasters. But the time and effort, not to mentions the billions of dollars, that giant multinational corporations have put into nuclear power will make it most difficult for them to relinquish the commitment to this energy source.

MICROWAVE ENERGY

IF THE RADIATION LEAKING FROM NUCLEAR WASTE DUMPS AND DAM-aged reactors doesn't kill you, homegrown radiation from ubiquitous cell phones may do the trick.

After all, a cell phone is merely a handheld radio transmitter/receiver both sending and receiving microwave energy. Anyone with a microwave oven knows what a high-powered blast of energy does to organic tissue. But what about low-level energy? The fact is that no one knows, since cell phones have only been around for about twenty years.

There's no question that cell phones have skyrocketed in popularity over that period. According to the Cellular Telecommunications and Internet Association, as of 2010, there were more than 303 million subscribers to cell-phone service just in the United States. This was almost triple the 110 million users in 2000. Worldwide, the number of cell-phone subscriptions is estimated to be 5 billion.

But new studies have increasingly indicated the deadly effects of prolonged cell-phone use. According to the results of a French study published In the British journal *Occupational and Environmental Medicine,* individuals who used their cell phones for more than fifteen hours per month over a five-year period on average had between two and three times greater risk of developing glioma and meningioma (brain and spinal tumors) compared with people who rarely used their phones.

This new study only added strength to past studies that had suggested a link between long-term cell-phone use and brain tumors. "Our study is part of that trend, but the results have to be confirmed," said Isabelle Baldi, of the University of Bordeaux in south-western France, who took part in the study.

The researchers did agree that due to the ever-increasing use of cell phones, microwave intensity has been decreasing, making it difficult to know just how harmful phone usage might be. However, the French findings correlate with a Swedish study in 2003 that showed increased risk for glioma with cumulative cell-phone use. This research showed an increased risk of tumors in persons who began using cell phones before the age of twenty. Apparently, the highest risk of brain cancer from cell-phone use comes is related both to extended use (more than fifteen hours a week) and to beginning cell-phone use at an early age.

Dr. Richard A. Stein, a postdoctoral research associate in the department of molecular biology at Princeton University, wrote that the question of cell-phone toxicity has polarized society as much as the decades-long controversy over the ill effects of tobacco. He noted that "cigarette ads continued for decades, featuring health professionals, babies, and even Santa. It took thousands of scientific studies until establishments that initially adamantly refuted any links admitted to the potential health dangers."

Stein observes that several inconclusive studies on the dangers of cell phones have been funded by the cell-phone industry. He also notes that while thermal or "heat" issues have been addressed by cell-phone studies, many other factors may be involved, such as protein leakage through the blood-brain barrier and other cellular damage. He notes some studies also demonstrate "significant correlations with infertility, decreased sperm counts, viability and mobility" in men who carry their cell phones below the waist. "The mere fact cell-phone booklets warn to keep the device at a certain distance from the body is one of the strongest indications that the radiation emitted is not totally harmless. As with every exposure, some individuals will be less sensitive, other will be more sensitive, but the mere warning shows something is going on," explains Stein.

The phone aside, new studies also indicate that just being near to a cell-phone tower can be hazardous. A 2013 article in the *British Medical Journal* (*BJM*) concluded that proximity (anywhere up to

1,600 feet) to cell-phone towers resulted in lack of concentration, vertigo, irritability, difficulty sleeping, and lack of appetite. There are currently more than 190,000 towers in the U.S., with more being added all the time.

Towers communicate with nearby phones through radiofrequency (RF) waves, a form of energy in the electromagnetic spectrum between FM [frequency modulation] radio waves and microwaves. RF waves are different from stronger types of radiation such as X-rays, gamma rays, and ultraviolet (UV) light, which can break the chemical bonds in DNA and at high levels can damage body tissues, as in microwave ovens. No one is certain what may result from ongoing low-level FM radiation.

Today, the public is exposed to one hundred million times more electromagnetic radiation than previous generations. If you can make a cell-phone call, you're in an area saturated with microwave radiation.

Electronic signals from cell-phone towers extend indefinitely but typically continue to decrease in intensity within a radius of twenty-one miles. Almost all Americans, with the exception of a few who live in isolated spots in the western half of the country, live within twenty-one miles of a tower, meaning a majority of the population is constantly exposed to cell-tower radiation. Cell towers, which broadcast and receive electromagnetic switching signals, have long been suspected of interfering with normal brain and body functioning, as human biology relies on electrobiochemical pathways for healthy function. The journal article added support to those who claim that electronic pollution, not just industrial pollution, may be harmful to healthy humans.

In the past, those persons claiming that cell-phone energy caused them neurological harm, termed "electromagnetic hypersensitivity," were dismissed by some doctors and industry scientists as delusional. However, the *BJM* study reported that the phenomenon is real, with eight of ten studies the scientists evaluated reporting "increased prevalence of adverse neurobehavioral symptoms or cancer in populations living at distances five hundred meters from BSs [cell-tower base stations]."

The author of the study noted that the cell towers they examined all met current safety guidelines, indicating that these guidelines are inadequate to safeguard the public from electromagnetic radiation. The study also found that since cell-tower signal strength is determined by the inverse square of the distance, a person who lives twice as close to a cell tower receives four times the radiation.

Close observers of our society have noted how the general public appears increasingly confused and irritable. *NaturalNews* editor Mike Adams wrote, "A society that once operated with some degree of sanity and politeness has become largely demented and rude. Mathematical abilities are nearly lost across the population, as very few people under the age of forty can even calculate 15 percent waiter's tips at a restaurant. The ability of voters to understand laws, liberties, freedom and even the structure of government is almost entirely lost in nations where cell-phone towers are ubiquitous. Given this recent research revealing the negative impact of cell-phone radiation on human brain function, it would be incredibly irresponsible to fail to consider how cell-tower radiation alters healthy brain function and promotes confusion and irritability."

Many thoughtful people in America have begun treating their cell phone like their credit card—useful in an emergency but to be used only on occasion. Parents should be especially mindful of the time their children spend on a cell phone.

CHEMTRAILS

A FEW YEARS BACK, A RETIRED ARMY INTELLIGENCE OFFICER WAS visiting a friend still in the military and stationed at the White Sands Proving Ground in New Mexico, when a high-flying jet flew across the sky leaving a white trail in its wake. During the conversation, the serviceman boasted about the security at White Sands, stating that no aircraft were allowed to fly over its restricted airspace. The former intelligence officer pointed to the craft above them and asked, "What about that airplane?" After looking directly at the jet above them, the serviceman replied with a smirk, "What plane?"

This anecdote illustrates the problem of the aerial phenomena that have come to be known as Chemtrails. Like the serviceman under orders not to admit to overflights, no one in a position of authority will admit that Chemtrails exist, much less who is responsible for them and what purpose they serve. But unlike many modern mysteries, this one is visible to anyone who cares to look up on the days that large high-flying jets weave narrow and continuous vapor/chemical trails through the sky. Despite dismissive articles calling Chemtrails merely a new "conspiracy theory" and citizens in denial that something so obvious might not be reported in the mainstream media, a growing number of serious researchers have studied the phenomenon.

"For more than a decade, first the United States and then Canada's citizens have been subjected to a 24/7/365-day aerosol assault over our heads made of a toxic brew of poisonous heavy metals, chemicals, and other dangerous ingredients. None of this was reported by any mainstream media," noted the late environmental activist and former college professor Dr. Ilya Sandra Perlingieri. "The U.S. Department of Defense [DOD] and military have been systematically blanketing all our skies with what are known as Chemtrails (also known as stratospheric aerosol geoengineering). These differ vastly from the usual plane contrails that evaporate rather quickly in the sky."

Chemtrails are nothing new. The U.S. military, in what have been described as "vulnerability tests," has sprayed chemical and biological compounds in open-air testing over civilian populations since the 1940s.

Just one example was the dumping of thousands of pounds of zinc cadmium sulfide during nearly three hundred secret experiments between 1952 and 1969. Targets of this contamination included Fort Wayne, Indiana; St. Louis; San Francisco; Corpus Christi; Oceanside, California; and even Minnesota's Chippewa National Forest.

Beginning around 1997, observant people around the world starting noticing the long trails in the sky that failed to evaporate like the condensation trails from normal aircraft. Condensation trails, or contrails, have been a fixture in the skies since World War II, when high-altitude bombers would leave a trail of condensation behind them.

Any aircraft engine, jet or piston, produces warm moist air, which, when injected into the cold dry air of the upper atmosphere, results in a trail of water vapor or ice crystals that stream out behind the craft. Once these particles return to a cooler state, they evaporate back into the air. This vaporization can take place within ten seconds or stretch for more than an hour depending on the temperature and humidity in the atmosphere. Since contrails primarily contain water, they present no real hazard to the population. Contrails do not normally occur below about thirty thousand feet.

Chemtrails are entirely different. These trails do not evaporate but spread out and eventually form a cloudy haze over the entire sky. As World War II veteran David Oglesby noted after observing Chemtrails in the sky above his California home, "The trails formed a grid pattern. Some stretched from horizon to horizon. Some began abruptly, and others ended abruptly. They hung in the air for an extended period of time and gradually widened into wispy clouds resembling spiderwebs. I counted at least eleven different trails."

The official debunking line that all observed trails in the sky are simply condensation trails falls apart when Chemtrails are seen simultaneously in the same portion of sky and at approximately the same altitude. Keen observers have noticed two planes of similar size laying trails at the same altitude and at the same time. Yet they are not the same. One is short, while the other reaches to the horizon. Others ask how it can be that one day there are numerous trails crisscrossing each other in the sky but on the next day, with no change in the climate, there are no trails in the sky. Did all aircraft suddenly stop flying?

Even more damning, Chemtrails often occur at altitudes and in conditions that would make it impossible for a normal contrail to form. Furthermore, contrails are composed of water vapor, whitish in color, and produce no "halo" effect in sunlight. Chemtrails, on the other hand, appear oily and produce rainbow effects, especially in the late afternoon sun. Contrails cannot be stopped without shutting off an aircraft's engines, but Chemtrails have been observed coming to an end even as the aircraft producing them flies onward. Aircraft do not shut off their engines midflight.

Ohio representative Dennis Kucinich, who served as chairman of a subcommittee of the House Committee on Oversight and Government Reform, once introduced his unsuccessful Space Preservation Act of 2001 (HR 2977), designed to ban the deployment of exotic weapons in space. Listed in this legislation were chemical, biological, environmental, climate-changing, or tectonic weapons and, notably, Chemtrails. In 2004, while campaigning as a presidential candidate, Kucinich responded to a question about Chemtrails by flatly stating, "Chemtrails are real."

Bob Fitrakis, of *Columbus Alive* asked Kucinich why he would introduce a bill banning so-called Chemtrails when the U.S. government routinely denies their existence and the U.S. Air Force routinely calls Chemtrail sightings hoaxes. Kucinich replied, "The truth is there's an entire program in the Department of Defense, 'Vision for 2020,' that's developing these weapons." In an apparent effort to conceal classified weapons systems, the more exotic weapons, including Chemtrails, were later stricken from the failed bill.

Government, aided by a compliant corporate mass media predisposed to blindly accept official pronouncements as gospel, continues to skirt the issue. No ambitious reporter wants to sidetrack their career by digging into the morass of official denials on Chemtrails. Even many politicians appear to be in the dark, having been given the official runaround when it comes to Chemtrails. After describing condensation vapor trails, and attributing Chemtrail sightings to ever-increasing air traffic, air-force colonel Walter M. Washabaugh testified in Congress that Chemtrails are a "hoax." He stated, "The Air Force's policy is to observe and forecast the weather to support military operations. The Air Force is not conducting any weather modification experiments or programs and has no plans to do so in the future. In short, there is no such thing as a 'Chemtrail'—the actual contrails are safe and are a natural phenomenon. They pose no health hazard of any kind."

A Houston, Texas, study found that over the course of sixty-three days of observation, military aircraft laid down white plumes that lasted for eight hours or more, while commercial airliners flying at the same altitude left contrails that lasted no longer than twenty-five seconds. Stranger still were observations of aircraft not filing flight plans and hence not recorded by the FAA or shown in the flight software. "It was discovered that the jets that were responsible for leaving highly persistent trails that last for hours did not ever appear on Flight Explorer," the report stated.

So how are Chemtrails harming us? In the late 1990s and early 2000s, people—and even some local TV stations—began to report strange white sticky substances being found on homes, yards, and vehicles. It was lacelike and hung from trees and shrubbery as if it had fallen from the sky. Chemical analysis found this substance contained aluminum oxide, barium, polymers, and even traces of pathogens. People in the vicinity of this substance reported ill effects such as asthma, fatigue, headaches, dizziness, joint pain, and flulike symptoms. The Chemtrail formula was apparently improved, as the number of reports of webby material has declined in recent years.

Because no one in a position of authority will admit what is happening right in front of the public's eyes, researchers have been forced to speculate on what Chemtrails are about.

One theory is that the citizenry is being exposed to chemical spraying as a covert means of inoculation, although germ warfare experts and medical authorities agree that a high-altitude spraying program is an inefficient method for distributing bacteria and viruses. Likewise, there seems to be little evidence that the Chemtrails are specifically designed to cull the human population, as was believed by many in the early days of the spraying.

Many pathogens have been found in the upper atmosphere, which is an indication that illnesses brought on by Chemtrails may not be an intentional part of the program. Nevertheless, they may be inadvertently brought to earth by the Chemtrail haze. This undoubtedly is a small consolation to those who suffer from such contamination, and indicates an indifference to public harm on the part of those responsible for the spraying.

One Canadian research foundation concluded that the much-discussed but little-publicized Chemtrails may be an attempt to hide a sickening military secret. Professor Donald Scott, president of the Common Cause Medical Research Foundation, posits that Chemtrails are a belated attempt by U.S. military and intelligence chieftains to stop the spread of a debilitating disease first concocted by the U.S. military in the early 1980s.

According to Scott's account, the military began developing diseases in the 1970s that were infectious but not contagious. In other words, they developed an ailment that could be spread to enemy troops but would not pass into other populations. One such disease was based on brucellosis, a disease that can be transmitted to humans by animals. Brucellosis is a bacterial disease usually found in cattle that can cause undulant fever in humans. By manipulating the disease, researchers were able to design a disabling bacteria that disappeared following infection. Troops could be infected yet exhibit no signs of the bacteria when examined by a doctor.

Scott goes on to say that, encouraged by successful testing and with the approval of then vice president George H. W. Bush, in 1986, the brucellosis bio agents were shipped to none other than Saddam Hussein in Iraq, who at that time was fighting a protracted war against Iran at the behest of the CIA. But after Saddam had obtained a stockpile of *Brucella abortus,* biotypes 3 and 9, and *Brucella melitensis,* biotypes 1 and 3, it was discovered that this designer bacteria had mutated and become contagious.

According to Scott's report, Saddam used this pathogen on American troops during the Persian Gulf War in 1991, resulting in the illness referred to as Gulf War syndrome. More than one hundred thousand Gulf War vets still suffer from this illness, which causes chronic fatigue, loss of appetite, profuse sweating even at rest, joint and muscle pain, insomnia, nausea, and damage to major organs. Much of this information may be found in a 1994 report by Senator Donald W. Riegle Jr. entitled, "U.S. Chemical and Biological Warfare–related Dual Use Exports to Iraq and Their Possible Impact on the Health Consequences of the Persian Gulf War." Some claim a variant of the brucellosis brought home by soldiers has spread to the civilian population in the U.S., with many people now suffering from general debilitation and tiredness.

When the contagion began to spread among the general population, top officials with the National Institutes of Health and Centers for Disease Control as well as the Defense Department and the Department of Health and Human Resources began a program of misrepresentation of the disease to mask their role in its origin. The illness, initially said to be psychosomatic, later was claimed to be connected to the Epstein-Barr virus and was labeled "chronic mononucleosis." This has now become known as chronic fatigue syndrome. Like the veterans before them, victims of this ailment were told it was merely a psychological condition.

Meanwhile, officials reportedly took steps to counteract the pathogen by covertly inoculating the public using airborne biological agents within Chemtrails.

Media outlets have been complicit in the government's denial of Chemtrails, and attempts to explore the phenomenon have been woefully inadequate. In February 2007, the Discovery Channel aired a program devoted to the issue. Many believed this was a classic case of misdirection. The program, *Best Evidence,* was entitled "Chemical Contrails" and focused on whether jet-aircraft vapor trails might be toxic. The first third of the program fairly presented the concerns of several citizens and researchers regarding the aerial trails that appeared much different from the usual vapor trails. The next third described various scientific equipment, including a jet engine, which was to be used to test burned particulates from the jet's exhaust. The final third was an exhaustingly detailed analysis of the burned fuel that showed no pathogens or any other harmful chemical or metal in the fuel waste.

The show made clear that nothing unusual was to be found in jet vapor trails. But the fatal flaw in its producers' analysis was obvious. The show's narrator stated that the U.S. military had refused to allow its jet fuel to be tested, which eliminated the prime suspect for Chemtrails. For their test, the show's producers had used commercial jet fuel purchased from a local airport. No wonder the burned fuel showed no unusual properties. No one has ever accused normal jet fuel of being contaminated with unusual chemicals or pathogens. It also has been established, both by observation and photographs, that Chemtrails emanate from lines along aircraft wings, not from the engines. This demonstrates that Chemtrails constitute an aerial spraying program, and are not connected to spent jet fuel.

One Louisiana TV station in late 2007 took upon itself the task of testing water captured under a crosshatch of aerial trails. KSLA-TV News in Shreveport found exceptionally high levels of barium, an alkaline earth metal rarely found in nature, since it quickly oxidizes in air. An excessive amount of barium is toxic, affecting the nervous system and the heart. State health officials confirmed to the TV station the danger of barium to the human nervous and immunization systems but hesitated to link this danger to Chemtrails.

Other tests had already identified barium, along with aluminum oxide, among the contents of Chemtrails. During a three-month period in 2002, three separate rainwater and snow samples from Chapel Hill, North Carolina, were collected and submitted for formal double-blind laboratory analysis. Therese Aigner, an accredited environmental engineer, found significant amounts of barium, aluminum, calcium, magnesium, and titanium in the samples, all of which had a verified chain of custody. Aigner concluded that the consistency of the findings indicated "a very controlled delivery (dispersion) of Chemtrails by aircraft in [the] area." She added that whoever was responsible for the Chemtrails was violating more than a half-dozen laws and regulations, both state and national.

The potential environmental effects of Chemtrails are especially alarming, after a five-year study around Mount Shasta in California determined that Chemtrails were causing eco-collapse. More than sixty-one thousand parts per billion (ppb) of aluminum were found in snow at Mount Shasta when the normal level should have been about 7 ppb. In addition, the soil pH had changed from acidic to neutral and the trees were dying. Most concerning was the fact that about 80 percent of amphibious species in tributaries had died off and the local potato harvest had declined by 80 percent.

In light of this fact, it may be more than coincidence that the chemical giant Monsanto has developed genetically modified aluminum-resistant seeds. Could it be that the environment is being poisoned deliberately in order to enable Monsanto to reap greater profits and gain even more control over the world's food supply?

Other researchers believe the purpose of Chemtrails may include: environmental modification and control, including weather control; biological operations; electromagnetic operations, including the High Frequency Active Auroral Research Program (HAARP) and cell towers; military applications; planetary and geophysical modification; a global surveillance system; and the detection of ionic disturbances from exotic propulsion systems. At least one researcher, noting that during the First Gulf War barium was fed to the Iraqi insurgents so they could be tracked by electromagnetic frequency devices, believes Chemtrails are monitoring individuals and groups.

These possible uses aside, there's clear evidence of the harm of another Chemtrail ingredient, tiny synthetic filaments called polymers. Polymer chemist Dr. R. Michael Castle reports that some of polymers he has studied in connection with Chemtrails can cause "serious skin lesions and diseases when absorbed into the skin." Such polymers may connect Chemtrails to another controversial issue— the strange and mysterious malady known as Morgellons disease that is reportedly infecting more than twelve thousand American families.

In addition to skyrocketing rates of lung cancer, asthma, and pulmonary/respiratory problems blamed at least partially on Chemtrails, Morgellons is an unexplained skin disorder that includes disfiguring sores and a crawling sensation both on the surface of the skin and under it. Strange fibers matching material from Chemtrails have been found on the sores of people suffering from the disease. But just as government officials deny the reality of Chemtrails, they also discount reports of Morgellons. CDC officials say they've found no evidence that Morgellons disease is caused by an infectious agent or a substance in the environment, and that those experiencing symptoms of Morgellons disease may actually be suffering from delusional parasitosis, a mental illness in which the patient believes they are infested with parasites.

Dr. Russell Blaylock says he initially treated reports of health problems stemming from Chemtrails skeptically, but has now reconsidered. "My major concern is that there is evidence that they are spraying tons of nanosized aluminum compounds. It has been demonstrated in the scientific and medical literature that nanosized particles are infinitely more reactive and induce intense inflammation in a number of tissues," he writes. "Of special concern is the effect of these nanoparticles on the brain and spinal cord, as a growing list of neurodegenerative diseases, including Alzheimer's dementia, Parkinson's disease, and Lou Gehrig's disease are strongly related to exposure to environmental aluminum."

One other possible ways elites may be using Chemtrails is to modify or control the weather. Experiments in this area date back into the 1950s, so it seems reasonable to assume that some advances have been made. In the 1990s, some scientists, including Dr. Edward Teller, the "father of the hydrogen bomb," and a founder of the "Star Wars" missile defense system, proposed seeding the upper atmosphere with millions of tons of sulfur or other heavy metals to create a cloud cover to deflect sun rays and prevent further heating of the earth. Some scientists warned such a program would turn blue skies milky white and perhaps cause droughts and further ozone depletion. Teller admitted the difficulty in persuading the public to allow a program that would pollute the air with metal particles, many known to be harmful to humans.

And in 1996, the air force published a research paper entitled "Weather as a Force Multiplier: Owning the Weather in 2025." The paper concludes that although "offensive weather-modification efforts would certainly be undertaken by U.S. forces with great caution and trepidation, it is clear that we cannot afford to allow an adversary to obtain an exclusive weather-modification capability." To this end, the paper states, "Efforts are already under way to create more comprehensive weather models primarily to improve forecasts, but researchers are also trying to influence the results of these models by adding small amounts of energy at just the right time and space." Anyone familiar with past secret government projects, such as the B-1 "Stealth" bomber, knows that if the military publicly speaks of a technology, even as a future possibility, it has already been developed.

Likewise, the globalists often telegraph their plans. A 2013 arti-

cle in *Foreign Affairs*, a publication of the Council on Foreign Relations (CFR), reiterates the belief in unusual environmental change,
stating, "It is clear that, unchecked, climate change won't just
menace natural ecosystems; it will also cause severe harm to humans
and could even threaten national security. And, because governments have made barely any progress in controlling the emissions
that cause global warming—the 2000s saw the most rapid growth
in emissions of carbon dioxide and other warming gases since the
1970s—it's not so crazy to imagine that some nation will launch an
emergency geoengineering scheme, perhaps before its viability and
consequences are understood." Chemtrails could be evidence of just
such a scheme. The article concludes, "Until the science gets serious,
the politics won't reflect what's really at stake. Meanwhile, the planet
keeps warming and the day when geoengineering might be needed
draws nearer."

This CFR publication openly discusses a strategy of reducing
harmful ultraviolet energy under the name "solar radiation management (SRM)." "The usual proposals involve spraying material into
the stratosphere, where it would turn into reflective clouds, or blowing seawater into the air, with a similar effect. The clouds could deflect just enough incoming sunlight to offset, crudely, the number of
degrees human emissions have warmed the planet," state the authors.
"Flying a fleet of high-altitude aircraft that spray particles into the
upper atmosphere would cost perhaps ten billion dollars per year—a
pittance for a country that is suffering from severe climate change
and seeks a quick solution."

President Obama's science adviser John P. Holdren in 2009 discussed "geoengineering" with administration officials and suggested
seeding the upper atmosphere with particles of "pollutants" to create
a heat shield in the hope of mitigating climate change. Holdren con-
ceded the possibility of "grave side effects," but said, "We might get
desperate enough to want to use it." Some researchers believe Holdren may have been testing the waters of public opinion with regard
to aerial spraying as with the Chemtrails.

An old journalist credo states, "Follow the money." If the world
continues to suffer the effects of global climate change, the losses

to the agriculture and insurance industries will be catastrophic. If flooding in the Mississippi Valley, drought and fires on the West Coast, and unusually strong hurricanes, tornadoes, and storms continue, the United States insurance industry could face bankruptcy, bringing down the entire national economy. Since so many foreign currencies are based on the U.S. dollar, this would precipitate a worldwide economic crisis. Apparently, someone feels the deaths of Chemtrail-susceptible people, whether the very young or old, is an acceptable cost to prevent this eventuality.

The injection of heavy-metal particles into the atmosphere via Chemtrails may support various military activities, such as enhancing radar and communications and even boosting the effects of the High Frequency Active Auroral Research Program (HAARP), a vast array of dishes transmitting powerful beams of electromagnetic energy into the upper atmosphere located near Gakona, Alaska. Officially, HAARP is designed to study the ionosphere, the uppermost portion of the earth's atmosphere. However, critics of the program claim this powerful tool may also be used as a weapon to deliver energy blasts equal to an atomic bomb, destroy communications across the planet, and even influence human behavior by broadcasting human brain-wave frequencies. Conspiracy theorists see the HAARP experiments as the possible cause of recent weather calamities such as hurricanes, tornadoes, floods, and earthquakes, with some websites regularly tracking HAARP frequencies.

As reported in the October 2004 edition of *Scientific American*, work has continued on ways to manipulate weather by using directed energy or seeding clouds with silver iodide from aircraft. Theoretically, directed energy, such as produced by HAARP, could be used to lessen the strength of a hurricane and, by heating adjoining sea water, actually guide a hurricane away from populated areas. Of course, if such techniques prove viable, hurricanes may be intensified and used as a weapon by guiding them into a specific target, as seen when Hurricane Katrina suddenly veered into New Orleans.

In June 2014, the air force announced to Congress that it was closing the Alaskan HAARP facility. David Walker, air-force deputy assistant secretary for science, technology, and engineering, told the

Senate, "We're moving on to other ways of managing the ionosphere, which the HAARP was really designed to do . . . to inject energy into the ionosphere to be able to actually control it. But that work has been completed." The website geoengineeringwatch.org lists many other known and suspected HAARP sites in other Alaska locations as well as in Colorado, Massachusetts, California, Texas, Norway, Germany, Wales, Peru, Australia, Japan, and Russia.

Another scary purpose behind the Chemtrails was voiced by a retired military intelligence officer who became particularly interested in the phenomenon after being doused with chemicals from the air that seriously affected his health. This man participated in the army's remote viewing program, so he was able to psychically seek the purpose behind the Chemtrail program. What he got was summed up in one word—"terraforming," an attempt to alter the environment of the planet.

The case for the reality of Chemtrails is strong, but until there is some official recognition of this activity, the tax-paying public will be left to do no more than speculate about its originators and purpose.

CHANGING THE GAME

EVERY LIVING THING REQUIRES AIR. CONTINUING TO PUMP POLLUT-ants into the air is suicidal. Every effort must be made on the individual level to maintain clean air, and such action should include ending any dependence on nuclear power, proven to be a hazardous and unpredictable technology.

Cell phones should be kept away from children and their use severely limited for adults. If a phone must be used, a headset should be worn to keep heat and radiation away from the brain.

The public must recognize the dangers of the aerial spraying program producing Chemtrails and demand a full and honest investigation by both the government and the news media. The public can no longer simply take the government's word that nothing is going on in the skies above them.

But once again, public policies on pollution, hazardous technologies, and environmental modification will not change as long as corporate and military power prevails in governments. A basic restructuring of political power must take place, first at the local level. Only then can effective efforts be made to ensure a safe and clean environment for everyone.

CHAPTER 15

A POLICE STATE

☐ UR ACCEPTANCE OF ALL OF THESE DEADLY PRODUCTS IS ENFORCED by draconian measures of surveillance and punishment. Life in the emerging police state of the American republic ranges from the manifestly ridiculous to the deadly.

"Whether it's the working mother arrested for letting her nine-year-old play unsupervised at a playground, the teenager forced to have his genitals photographed by police, the underage burglar sentenced to twenty-three years for shooting a retired police dog, or the forty-three-year-old man who died of a heart attack after being put in a chokehold by NYPD officers allegedly over the sale of untaxed cigarettes, the theater of the absurd that passes for life in the American police state grows more tragic and incomprehensible by the day," comments John W. Whitehead, founder of the civil liberties watch group the Rutherford Institute.

Journalist Chris Hedges is an activist Presbyterian minister and best-selling author of several books as well as a finalist for the National Book Critics Circle Award for nonfiction, senior fellow at New York City's Nation Institute, and columnist for the websites Truthdig and OpEdNews.

He noted that the fall of the Soviet Union left America without a competing superpower, which threatened to delegitimize its massive spending on war and state security, today totaling more than 50 percent of the national budget. Currently it is a group of Islamic radicals who have taken the place of the old communist bloc. The fear and the psychosis of permanent war continue while the "War on Terrorism" has created new and more complicated demands on our intelligence agencies. "Our illegal and disastrous occupations of Iraq and Afghanistan and our indiscriminate bombing of other countries, along with the war crimes Israel is carrying out against the Palestinian people, are driving people in the Muslim world into the arms of these militant groups. We are the most hated nation on earth," wrote Hedges.

"At the same time, globalization—our corporate policy of creating a worldwide neo-feudalism of masters and serfs—means we must spy on citizens to prevent agitation and revolt. After all, if you are a worker, things are only going to get worse. To quash competitors of American companies, we spy on corporations in Brazil, including Brazil's biggest oil company, Petrobras, and on corporations in Germany and France. We also steal information from the leaders of many countries, including German chancellor Angela Merkel, whose personal cell phone we tapped. However, Ms. Merkel, who grew up in East Germany, should not, as she has done, accuse us of being the Stasi. We are much more efficient than the Stasi was. We spied successfully on UN Secretary-General Ban Ki-moon, in addition to Pope Francis and the conclave that elected him last March. Senior UN officials and Roman Catholic cardinals are highly susceptible to recruitment by al-Qaida . . . Threats to the nation raised new legal and policy questions, which fortunately our courts, abject tools of the corporate state, solved by making lawful everything from torture to wholesale surveillance."

Today, along with being accused of trying to play policeman to the world, U.S. rulers clearly appears to be attempting to turn the United States republic into an Orwellian police state. And such a state begins with ubiquitous spying on its own citizens.

A SURVEILLANCE STATE

EVEN DISREGARDING THE MANY EDUCATED AND THOUGHTFUL
people who have come to see the attacks of September 11, 2001, as an
inside job perpetrated by elements within the U.S. government, it is
now clear that political and military leaders used the horrific attacks
as a pretext to raise up both a national security surveillance state and
corporate defense profits.

In post-9/11 America, government intrusion into our lives has
grown to the point where the presumption of innocence no longer
applies. In July 2014, new revelations of National Security Agency
(NSA) spying showed that the spy agency was targeting even prominent American citizens with no criminal records. American lawyer,
journalist, and author Glenn Greenwald, who worked as a columnist in U.S. offices of the *Guardian,* along with journalist and political commentator Murtaza Hussain, whose work has appeared in
the *New York Times,* the *Guardian,* the *Globe and Mail,* and *Salon,*
among others, produced government documents obtained from
whistle-blower Edward Snowden showing how the NSA and the FBI
"covertly monitored the emails of prominent Muslim-Americans—
including a political candidate and several civil rights activists, academics, and lawyers—under secretive procedures intended to target
terrorists and foreign spies."

Faisal Gill, one of those targeted by government surveillance,
said he is a longtime Republican Party operative who once served in
the Department of Homeland Security under President George W.
Bush and held a top-secret security clearance.

"I went to school here as a fourth grader—learned about the
Revolutionary War, learned about individual rights, Thomas Jefferson, all these things," said Gill. "That is ingrained in you—your privacy is important. And to have that basically invaded for no reason
whatsoever—for the fact that I didn't do anything—I think that's
troubling. And I think that certainly goes to show how we need to
shape policy differently than it is right now."

Mass surveillance by the U.S. can be traced back to the Spanish-

American War when U.S. forces in the Philippines maintained card files on 70 percent of the population. This system was brought back to the U.S. and incorporated into legislation such as the 1917 Espionage Act and the 1918 Sedition Act.

During World War II, the FBI set up surveillance systems in foreign countries such as Brazil and Colombia. After the war, the espionage territory was expanded to Japan, Greece, Uruguay, and other countries feared susceptible to communist infiltration. In Guatemala in 1954, the CIA used various methods of surveillance in its effective overthrow of the government of Jacobo Árbenz. American advisers then helped create a vast filing system there to hunt down enemies of the new U.S.-backed regime. This program, known as the National Police Archive, tracked the movements of dissidents, recorded their political opinions, identified their associates, mapped their daily routes, and, ultimately, eliminated them.

Today, the cutting edge of the surveillance state is the NSA. Prior to the 2013 Snowden leaks, few people were aware of the NSA's existence, even as it rapidly grew around them. Those who were aware of the NSA's existence but were under restrictions not to talk about it laughingly referred to it as "No Such Agency."

Snowden, a former NSA contractor, was given sanctuary in Russia after U.S. authorities sought his arrest for exposing mass global surveillance programs led by the NSA and Government Communications Headquarters (GCHQ), its British counterpart. The leaks revealed these organizations' disturbing practices, which included tapping Internet networks, emails, and phone calls of millions of ordinary citizens, as well as the political leaders of other nations.

Though branded a traitor by hard-line war supporters and intelligence chiefs at home, Snowden provided a thought-provoking explanation for his actions in 2012 with a statement not widely reported in the corporate mass media: "A little over one month ago, I had family, a home in paradise, and I lived in great comfort. I also had the capability without any warrant to search for, seize, and read your communications. Anyone's communications at any time. That is the power to change people's fates.

"It is also a serious violation of the law. The Fourth and Fifth

Amendments to the Constitution of my country, Article 12 of the Universal Declaration of Human Rights, and numerous statutes and treaties forbid such systems of massive, pervasive surveillance. While the U.S. Constitution marks these programs as illegal, my government argues that secret court rulings, which the world is not permitted to see, somehow legitimize an illegal affair. These rulings simply corrupt the most basic notion of justice—that it must be seen to be done. The immoral cannot be made moral through the use of secret law.

"I believe in the principle declared at Nuremberg in 1945: Individuals have international duties which transcend the national obligations of obedience. Therefore individual citizens have the duty to violate domestic laws to prevent crimes against peace and humanity from occurring.

"Accordingly, I did what I believed right and began a campaign to correct this wrongdoing. I did not seek to enrich myself. I did not seek to sell U.S. secrets. I did not partner with any foreign government to guarantee my safety. Instead, I took what I knew to the public, so what affects all of us can be discussed by all of us in the light of day, and I asked the world for justice.

"That moral decision to tell the public about spying that affects all of us has been costly, but it was the right thing to do and I have no regrets."

Snowden's actions reflect the fact that today the U.S. surveillance state has expanded to an unimaginable extent.

Christopher Ketchum, writing in *Radar* magazine, quoted a senior government official who carried high-level security clearances through five administrations and who revealed that "there exists a database of Americans who, often for the slightest and most trivial reason, are considered unfriendly, and who, in a time of panic, might be incarcerated. The database can identify and locate perceived 'enemies of the state' almost instantaneously."

Ketchum reports that the database is sometimes referred to by the code name "Main Core" and that some eight million Americans were listed in Main Core as potentially suspect in 2008. It is suspected that many more have been added since then. "In the event of a national emergency, these people could be subject to everything from heightened surveillance and tracking to direct questioning and possibly even detention," writes Ketchum.

Reportedly, the Main Core database, begun in the early 1980s, collects and stores—without warrants or court orders—the names of and detailed data about Americans considered to be threats to national security. Tim Shorrock, in *Salon* magazine, wrote, "According to several former U.S. government officials with extensive knowledge of intelligence operations, Main Core in its current incarnation apparently contains a vast amount of personal data on Americans, including NSA intercepts of bank and credit card transactions and the results of surveillance efforts by the FBI, the CIA and other agencies.

"One former intelligence official described Main Core as 'an emergency internal security database system' designed for use by the military in the event of a national catastrophe, a suspension of the Constitution or the imposition of martial law. Its name, he says, is derived from the fact that it contains 'copies of the "main core" or essence of each item of intelligence information on Americans produced by the FBI and the other agencies of the U.S. intelligence community.'"

Assistant professor of history at Harvard University Kirsten Weld noted, "The National Security Agency's surveillance leviathan, funded by a black budget and presided over by a star-chamber court, suctions up almost inconceivable amounts of material from around the world, including your phone and computer."

Weld wrote, "As one former U.S. intelligence official explained, 'rather than look for a single needle in the haystack'—scanning for information on particular cases of interest—the new strategy is now to 'collect the whole haystack.' This began in earnest with the Real Time Regional Gateway program [RTRG], implemented in Iraq and then in Afghanistan to vacuum up all possible information. The ethos of RTRG appeared in the U.S. in the form of the PRISM data-mining program. Americans were scandalized to learn from . . . Edward Snowden that the whole haystack included their phone calls and emails. They should understand that this will remain the case for as long as the U.S. is permitted to maintain its amorphous campaign against 'terror,' the diffuse goals of which are now seen to require a blanket approach to information gathering."

A centralized governmental database carrying files on all citizens, regardless of any criminal or mental health record, makes a mockery out of any concept of individual freedom.

One of the most egregious examples of such spying came in mid-2014 when Snowden, in an interview, revealed that in "numerous instances" he saw NSA employees passing around nude photographs intercepted "in the course of their daily work."

"You've got young enlisted guys, eighteen to twenty-two years old," Snowden explained. "They've suddenly been thrust into a position of extraordinary responsibility where they now have access to all of your private records. In the course of their daily work they stumble across something that is completely unrelated to their work in any sort of necessary sense. For example, an intimate nude photo of someone in a sexually compromising position. But they're extremely attractive.

"So what do they do? They turn around in their chair and show their co-worker. The co-worker says, 'Hey that's great. Send that to Bill down the way.' And then Bill sends it to George and George sends it to Tom. And sooner or later this person's whole life has been seen by all of these other people. It's never reported. Nobody ever knows about it because the auditing of these systems is incredibly weak. The fact that your private images, records of your private lives, records of your intimate moments have been taken from your private communications stream from the intended recipient and given to the government without any specific authorization without any specific need is itself a violation of your rights. Why is that in a government database?"

A spokesperson for the NSA, while not specifically denying Snowden's allegation, did say the agency has zero tolerance for such violations of the agency's standards. However, a September 2013 letter from the NSA's inspector general Dr. George Ellard to Iowa senator Chuck Grassley mentioned instances in which NSA agents admitted spying on former love interests.

In early March 2015, the House approved the Department of

Homeland Security's request for a yearly budget of almost $40 billion. However, the truth of the matter is that the system of national security is so vast and so secret that no one in the public has any idea of the size of some programs or how much they truly spend.

New technologies and the expansion of America's two dozen alphabet intelligence agencies has created a global corporate system of surveillance infringing on the rights of people both at home and abroad. The sometimes violent reaction of foreign populations to U.S. spying and military adventures only ensures endless conflicts and wars. But this also ensures endless profits for the war industries owned by globalists, which may be the real objective.

Such massive surveillance concerns many citizens on both ends of the political spectrum. What has been touted by government officials as a means to protect us has slowly been transformed into a surveillance state being used to protect the government and the people who control it from any public rebellion. As the major mainstream media today is under the domination of a mere handful of multinational corporations, there will be little investigation of the potential dangers of surveillance, and what reporting there is will likely be mere assurances that spy programs are for our protection. Critics are concerned that such surveillance may be used to identify and round up dissidents when government policies and the economy reach a breaking point.

Ironically enough, such intrusive surveillance and security methods apparently apply only to honest, tax-paying Americans, not to lawbreakers.

In mid-2014, the National Border Patrol Council (NBPC) revealed that many of the thousands of illegal immigrants crossing the Texas-Mexico border were being allowed to bypass TSA scrutiny and board and fly commercial airliners and buses by simply presenting an easily faked government document.

Illegals caught entering the USA are issued an "Application for Asylum and for Withholding of Removal" form (I-862) from the U.S. Citizenship and Immigration Services (USCIS), referring them to an immigration judge at some future date. This form can be used for identification purposes. "This just adds insult to injury," complained Hector Garza, a spokesman for NBPC's Local 2455. "Not only are we releasing unknown illegal aliens onto American streets, but we are allowing them to travel commercially using paperwork that could easily be reproduced or manipulated on any home computer . . . The [I-862] Notice to Appear form has no photo, anyone can make one and manipulate one . . . We do not know who these people are, we often have to solely rely on who they say they are, where they say they came from, and the history they say they have. We know nothing about most of them."

Modern America can truly be described as a nation more and more closely matching the nightmarish Big Brother world of Orwell, complete with cameras and listening devices in both public places and private homes. Personal privacy appears to be a thing of the past, as more than a year after Snowden revealed the controversial practice of monitoring telephone calls, U.S. intelligence agencies continue attempting to extend their bulk collection of American telephone records. In June 2014, the secretive Foreign Intelligence Surveillance Court (FISA) for the fifth time approved a request from the NSA that allowed the agency a ninety-day extension on its collection of metadata.

James Clapper, director of national intelligence, said the extension was necessary because the congressional reform process supported by Obama was not yet complete. Many researchers expressed doubt that such reforms would substantially alter government spying.

Well into 2014, President Obama was still formulating plans to overhaul NSA's telephone surveillance program. However, some U.S. officials claimed that the changes envisioned could actually compel telecommunications companies to collect and store customer data not previously required by law.

The Obama administration's plan would require carriers—not the NSA—to collect and store phone metadata, including dialed numbers and call lengths, though not the actual content of conversations. This could force companies to create new mechanisms to ensure that metadata from flat-rate subscribers could be monitored. According to one industry source, "These are very complex systems.

I doubt there are companies out there that have a nice, neat, single database that can tell you how long records are kept universally."

Verizon Communications' general counsel Randal Milch in a blog, while applauding proposals to end Section 215 (of the PATRIOT Act) bulk collection, argued that "the reformed collection process should not require companies to store data for longer than, or in formats that differ from, what they already do for business purposes." In other words, keep the status quo.

Amid such concerns, yet another NSA whistle-blower spoke at a July 2014 conference organized by the Centre of Investigative Journalism in London and revealed the NSA's ultimate goal. William Binney, a former NSA code breaker during the Cold War, resigned shortly after the 9/11 attacks in disgust over the agency's mass surveillance activities. Binney is said to be one of the highest-ranking whistle-blowers to come out of the NSA.

Binney said that while he is encouraged by recent Supreme Court decisions, such as the one requiring law enforcement officials to obtain a search warrant before downloading cell-phone data, "the ultimate goal of the NSA is total population control."

"At least 80 percent of fiber-optic cables globally go via the U.S.," he reported. "This is no accident and allows the U.S. to view all communication coming in. At least 80 percent of all audio calls, not just metadata, are recorded and stored in the U.S. The NSA lies about what it stores. The NSA is mass-collecting on everyone and it's said to be about terrorism, but inside the U.S. it has stopped zero attacks," he added.

Binney also expressed concern over the lack of oversight by the secret Foreign Intelligence Surveillance Court (FISA). "The FISA court has only the government's point of view," he noted. "There are no other views for the judges to consider. There have been at least fifteen to twenty trillion constitutional violations for U.S. domestic audiences and you can double that globally."

And the violations have expanded from collection of technological records into other facets of life. In early 2014, the Department of Homeland Security (DHS) was poised to activate a national license-plate tracking system shared with local law enforcement that would allow DHS officers to take smartphone photos of any license. These images then could uploaded to a national database that would include a "hot list" of "target vehicles." Similar surveillance systems in Britain and Australia have drawn criticism from civil libertarians. David Jancik, writing in Australia's *Courier-Mail,* comments, "The increasing use of number plate recognition technology by police opens a Pandora's box for abuse of power, mistakes and illegal disclosure. In a society that values civil liberties, this is absolutely unacceptable."

Many citizens are seriously concerned about the possibility of unauthorized abuse of such a system. "Do not kid yourself," warned James Smith with the Prepper Podcast Radio Network. "This is tracking of an individual that can be accessed at a whim. Yearly, officers are terminated for accessing the LEDS/NCIC database for looking into the histories of ex-lovers, future spouses, and potential son-/daughter-in-laws . . . And with a license plate tracking toy (not a tool), they will know where you are, as long as you have driven into the cross hairs of this new weapon for tyranny."

And the surveillance state has expanded into areas that once were confined to the realm of science fiction. In September 2014, the FBI announced that its Next Generation Identification (NGI) facial recognition system was "fully operational," ahead of schedule, despite claims the system identifies the wrong individual about 20 percent of the time. Minnesota Democratic senator Al Franken warned that the NGI system "could be abused to not only identify protesters at political events and rallies, but to target them for selective jailing and prosecution." A spokesman for the FBI said the NGI system, used by law enforcement agencies across the nation, was not expected to provide accurate information on a suspect but merely return a list of possible "candidates" as investigative leads. In response to a Freedom of Information Act request by the Electronic Frontier Foundation, the FBI provided documents revealing that its NGI system may include images of as many as fifty-two million Americans' faces. By the end of 2015, the FBI's entire biometric database may contain fingerprint or facial information on more than a third of the U.S. population.

Under such an inaccurate system, a person could be arrested, taken off the street, and held for some time because merely because they look like a wanted criminal.

Already, state and local law enforcement agencies use a wide variety of facial imagery databases to identify suspects. These include driver's licenses and Facebook photos.

The spread of such police state technology moved ahead in early 2015, as the Los Angeles County Sheriff's Department—the fourth largest police agency in the country—signed a $24 million contract to create a biometric database second only to that of the FBI.

Officials with Texas-based NEC Corporation of America, which provides biometric services to corporations, law enforcement, and governments across the world—said its six-year Los Angeles contract calls for providing the agency with specialized, state-of-the-art policing services, to include high-tech fingerprint and facial recognition software.

The new biometric service will interface with databases maintained by outside agencies including the FBI's NGI system, according to NEC officials. Although sheriff's officers have assured the public that no biometric data will be collected on innocent Los Angeles citizens, only on those arrested and charged with a crime, it was noted that criminal charges do not always lead to a conviction. Civil libertarians are concerned that millions of records on noncriminal Californians might potentially end up in the database.

The U.S. State Department is said to hold the largest facial imagery database in the federal government, storing hundreds of millions of photographs of American passport holders and foreign visa applicants. The Department of Homeland Security funds pilot projects in local police departments that match suspects' faces against those in crowd photographs.

According to NSA documents obtained in 2011 from former contractor Edward Snowden, the agency intercepts about fifty-five thousand "facial recognition quality images" out of millions captured each day. Agency officials termed this photo cache a "tremendous untapped potential" for cataloging every citizen. This potential was revealed by the *Guardian*, which disclosed that the NSA, along with its British counterpart, General Communications Headquarters (GCHQ), had intercepted webcam images from Yahoo subscribers, including sexually explicit material.

Such technology is still evolving, but already suspects can be identified even after growing a beard, changing locations, or trying to disappear within a crowd. Those concerned with civil liberties have expressed grave doubts about future privacy because of this constantly improving technology. Alessandro Acquisti, a researcher on facial recognition technology at Carnegie Mellon University, warned, "There are still technical limitations on it, but the computational power keeps growing, and the databases keep growing, and the algorithms keep improving."

Legislators have been slow to address this issue. Senator Franken, a former chairman of the Senate Subcommittee on Privacy, Technology, and the Law, in a 2013 letter to the head of the National Telecommunications and Information Administration (NTIA), remarked, "Unfortunately, our privacy laws provide no express protections for facial recognition data." The NTIA has contemplated standards and regulation for commercial use but not when it comes to the federal government.

The Next Generation Information system combines many different types of data in an individual's file, including such personal and biographic data as name, home address, Social Security (i.e., your computer number), immigration status, age, race, etc. This vast database is shared with the approximately eighteen thousand state, local, and tribal law enforcement agencies across the United States as well as other federal agencies. Unlike in the past, the NGI system will link both criminal and noncriminal databases.

As this database continues to grow, so also will its use. It is anticipated that soon employers requiring fingerprinting or a background check for job applicants will turn to the NGI system. "One of our biggest concerns about NGI has been the fact that it will include noncriminal as well as criminal face images. We now know that FBI projects that by 2015, the database will include 4.3 million images taken for noncriminal purposes," states Jennifer Lynch, a senior Electronic Frontier Foundation staff attorney.

When it was found that the resolution of many of the facial recognition photos was unacceptably low, FBI officials were quick to point out that its database is for investigative leads, not identification. This position prompted Lynch to ask, "It is unclear what happens when [a suspect] does not exist in the gallery—does NGI still return possible matches? Could those people then be subject to criminal investigation for no other reason than that a computer thought their face was mathematically similar to a suspect's?"

It should be noted that the company behind NGI's facial recognition database is MorphoTrust, a subsidiary of Safran, the French multinational firm dealing in aviation and aerospace products as well as security. MorphoTrust has placed facial recognition systems in approximately thirty-five state motor vehicle departments and in the Departments of Defense and State, which share records with the FBI.

The potential for mistakes in this giant web of databases, which now includes local jurisdictions such as the Los Angeles Sheriff's Office, is an uncomfortable reality for many.

And, as if checking license plates and citizens' faces with readers and street cameras were not enough, the government in early 2014 announced the deployment of the U.S. Army's Joint Land Attack Cruise Missile Defense Elevated Netted Sensor System (JLENS), a tethered blimp capable of detecting airborne objects up to 340 miles away and surface vehicles up to 140 miles away. According to one report, the blimp can view most of the Eastern Seaboard, from as far south as Richmond, as far west as Cumberland, Maryland, and as far north as Staten Island. Such airborne devices, now undergoing trials in the U.S., have been used for surveillance in Iraq, in Afghanistan, and along the U.S.-Mexico border.

ACLU privacy expert Jay Stanley told the *Washington Post* that the spy blimps represent "the kind of massive persistent surveillance we've always been concerned about with drones. It's part of this trend we've seen since 9/11, which is the turning inward of all of these surveillance technologies." In the past, defense systems such as the Distant Early Warning (DEW) line were aimed toward Russia. Today, they seem to be guarding against internal threats rather than foreign ones. Army spokespersons attempted to assuage any concern over the blimp monitoring by stating, "The primary mission . . . is to track airborne objects. Its secondary mission is to track surface moving objects such as vehicles or boats. The capability to track surface objects does not extend to individual people."

However the EFF's Lynch notes, "If it's able to track vehicles, that is problematic. You could imagine a scenario in which the location information can reveal where you go to church, what doctor you're going to, whether you're cheating on your wife, all those types of details . . . Once a surveillance technology is put up, it's very tempting for law enforcement or the military to use it for reasons they did not originally disclose."

The surveillance state now reaches into every aspect of our lives, including social media. Many of the more than one billion Facebook users would be horrified to learn that the NSA has co-opted the website in order to monitor citizens who have not been accused of any crime. According to internal government documents leaked by Snowden and reported by journalists Ryan Gallagher and Glenn Greenwald, "In some cases the NSA has masqueraded as a fake Facebook server, using the social media site as a launching pad to infect a target's computer and exfiltrate files from a hard drive. In others, it has sent out spam emails laced with malware, which can be tailored to covertly record audio from a computer's microphone and take snapshots with its webcam.

The hacking systems have also enabled the NSA to launch cyber attacks by corrupting and disrupting file downloads or denying access to websites. Gallagher and Greenwald, after studying NSA documents made available by Snowden, reported that before the NSA can extract data from a target, it must install malware on a targeted computer or network. "According to one top-secret document from 2012, the agency can deploy malware by sending out spam emails that trick targets into clicking a malicious link," they wrote. "Once activated, a 'backdoor implant' infects their computers within eight seconds. There's only one problem with this tactic, codenamed WILLOWVIXEN: According to the documents, the spam method has become less successful in recent years, as Internet users have become wary of unsolicited emails and less likely to click on anything that looks suspicious," they wrote. "Consequently, the NSA has turned to new and more advanced hacking techniques. These include performing so-called man-in-the-middle and man-on-the-side attacks, which covertly force a user's Internet browser to route to NSA computer servers that try to infect them with an implant."

The technical capability of the government, and the nefarious ends to which it is being put, is enough to drive the average person to the outer limits of paranoia. Government officials argue that these are unfortunate but necessity tools to fight terrorism. Yet the written evidence left by a number of NSA employees tells a vastly different story.

One unnamed official in the NSA's Signals Intelligence Directorate (SIGINT) posted his thoughts on a NSA discussion board that were included in a document released by Snowden. He wrote: "So, SIGINT is down right [sic] cool. As much as we complain about our 'Big Data Problem,' collection/processing issues, dismal infrastructure/outdated browsers/OS's [operating systems], our ability to pull bits out of random places of the internet, bring them back to the mother-base to evaluate and build intelligence off of them is just plain awesome!

"One of the coolest things about it is *how much* [emphasis in the original] data we have at our fingertips. If we 'only' collected the data we knew we wanted . . . yeah, we'd fill some of our requirements, but this is a whole world of possibilities we'd be missing! It would be like going on a road-trip, but wearing a blindfold the entire time, and only removing it when you're at one of your destinations . . . yeah, you'll see stuff, but you'll be missing out on the entire journey."

In another post, this government official admitted that the NSA "hunts" system administrators who control computer networks and servers, then targets their private email and Facebook accounts to gather information useful for hacking into their computers and gaining access to the entire network of computers they control.

And for all the liberties we have lost, this mass surveillance program is not even particularly effective. A 2014 study of terrorism cases by the New America Foundation showed that the NSA massive phone record collection has been virtually useless in preventing terrorist attacks.

The report states, "An in-depth analysis of 225 individuals recruited by al-Qaeda or a like-minded group or inspired by al-Qaeda's ideology, and charged in the United States with an act of terrorism since 9/11, demonstrates that traditional investigative methods, such as the use of informants, tips from local communities, and targeted intelligence operations, provided the initial impetus for investigations in the majority of cases, while the contribution of NSA's bulk surveillance programs to these cases was minimal. Indeed, the controversial bulk collection of American telephone metadata, which includes the telephone numbers that originate and receive calls, as well as the time and date of those calls but not their content, under Section 215 of the USA PATRIOT Act, appears to have played an identifiable role in initiating, at most, 1.8 percent of these cases."

In Europe, privacy advocates have had somewhat more success in curtailing surveillance by national governments. On April 8, 2014, the European Union Court of Justice struck down an online data collection directive issued by the EU claiming it "interferes in a particularly serious manner with the fundamental rights to respect for private life and to the protection of personal data." This was a blow to the intelligence agencies of both the UK and the U.S., which have operated under a 2006 Data Retention Directive requiring telecommunications companies to keep citizens' data up to two years. It was claimed such retention of data is "a necessary and effective investigative tool for law enforcement in several Member States, and in particular concerning serious matters such as organized crime and terrorism, it is necessary to ensure that retained data are made available to law enforcement authorities for a certain period."

Privacy International, a London-based group dedicated to fighting for the privacy rights around the world, issued a statement saying, "It is right and overdue that this terrible directive was struck down.

"This ruling not only demolishes communications data surveillance laws across Europe, but sets a precedent for the world. The widespread and indiscriminate collection of information has been, and always will be, bad law, inconsistent with human rights and democratic values. What the [whistle-blower Edward] Snowden revelations have showed us over the past year is that the international surveillance apparatus set up by intelligence agencies is in direct conflict with human rights. If the Data Retention Directive fails to meet the requirements of human rights law, then the mass surveillance programs operated by the U.S. and UK governments must equally be in conflict with the right to privacy . . . the international surveillance apparatus set up by intelligence agencies is in direct conflict with human rights."

Yet the EU still has a long way to go. It's telling that even Angela Merkel, chancellor of Germany, a close U.S. ally, in 2014 was denied access to her NSA file. And the Obama administration, despite assurances in late 2013 that the U.S. was no longer monitoring Merkel's phone calls, stated it would not answer formal questions from the Germans regarding NSA surveillance. Obama's government also refused to enter into a mutual "no-spy" agreement with Germany, saying Berlin was unwilling or unable to share the kinds of surveillance material the Americans requested.

In Europe, as in the United States, the most successful resistance has been at the individual level, as a few average citizens have begun to fight the "Big Brother" tactics of the surveillance state. In early 2014, a thirteen-year-old student in Wales fought back against increasingly common school surveillance by donning a mask representing Guy Fawkes, the English revolutionary convicted in the Gunpowder Plot of 1605, and now the most visible symbol of a collection of dissenters known as Anonymous. The girl refused to be fingerprinted as her school demanded. When a school employee tried to place her hand on a scanner, she withdrew it and stated that neither she nor her parents had given their consent.

Her mother told *Digital Journal:* "Of all the children in Melody's class only a few refused to be fingerprinted; the majority signed. I personally find this alarming. These are children not cattle, not convicted criminals—innocent children whose prints are now in the system. Anonymous unites people from every religion and country; never have I seen anything bring people together the way this does, people from all over suddenly put aside the feuds created by our governments, and are friends. There becomes a sense of being human and caring for one another that you simply do not get on such a wide scale; they are out there feeding the homeless, or fighting for justice for the people who have had their voices taken away forever. You can either be part of the solution or remain part of the problem."

Melody's case illustrates a broader issue in British schools. Despite a clause in the UK's Protection of Freedoms Act of 2012 guaranteeing parents that their permission would be asked before biometric data was taken from a child, it was estimated that roughly a third of Britain's schoolchildren have been duped into submitting to fingerprinting without parental consent. The civil liberties group Big Brother Watch submitted Freedom of Information requests to more than three thousand British schools and found about 40 percent were using this biometrical technology.

More fuel was added to this fire of outrage when *Sky News* reported that a class of three-year-olds in England was cajoled into giving fingerprints by being told it was a "spy game." One official defended the practice by claiming the fingerprints were intended for use in attendance reporting, along with verifying access to meals and libraries. It was said that prints were destroyed once pupils left the school.

Some have expressed the fear that the UK has been used as a testing ground for the global elite. After the successful introduction of fingerprinting for schoolchildren in Britain, the practice spread to the U.S. In 1999, Eagan High School in Minnesota encouraged students to use fingerprint readers to speed up the borrowing of library books. This was followed the next year by a school district in Pennsylvania where a local software development company designed a system whereby students bought lunch with just a fingerprint.

Children, with no real knowledge of history or privacy, are naively accepting of such new technology. "You don't have to bring lunch money. So, somebody can't take it," exulted one thirteen-year-old student.

Oddly enough, while the practice of fingerprinting schoolchildren has continued to grow in the UK and America, it has stopped in China, a nation long vilified in the Western media as a communist dictatorship with no respect for individual liberties or privacy. In 2006, Roderick Woo, justice of the peace at the Hong Kong Office of the Privacy Commission, ordered some schools to stop finger-printing children. The schools in Kowloon District were ordered to destroy all fingerprint data from students.

Woo explained, "It was a contravention of our law, which is very similar to your law, which is that the function of the school is not to collect data in this manner, that it was excessive and that there was a less privacy-intrusive method to use." He asked what better way is there for a teacher to take roll than to look around the class, note who's missing, and take down their names for the record. "Measuring fingerprints seemed a little over the top for the task in hand," he added.

Yet in America, allegedly the land of the free, such policies persist. To be fair, we have recently seen a few stirrings of resistance. In Houston, for example, when the Metropolitan Transit Authority (MTA) announced a plan to invite the Transportation Security Administration (TSA) to randomly search bags and packages, local citizens and libertarian attorneys were quick to offer opposition. After heated citizen input—Derrick Broze told the MTA board, "I don't feel like by purchasing a ticket or riding a bus [and having] to forfeit my constitutional rights and my protections and be subject to search or seizure. We don't plan on letting this issue die if the TSA stays in our city." Board chairman Gilbert Garcia said the press release was in error and that his agency had not conducted random bag searches on its buses and trains and did not plan to do so. However, Garcia later modified his statement, saying MTA officers might search bags but only based on probable cause or with a rider's consent. The episode illustrated that citizen protests in this arena can, occasionally, make a difference.

Similar security intrusion took place when the Massachusetts Bay Transportation Authority (MBTA) announced that all Boston bus and subway riders must submit to a warrantless search prior to boarding. Signage at a Boston "Transit Watch" checkpoint stated that "all persons choosing to use the MBTA transit system will be subject to security inspections of their handbags, briefcases and/or other carry-on items." The notice went on to state that anyone refusing a security inspection would be requested to leave MBTA property and/or be subject to arrest for trespassing.

Yet despite a few citizen protests, the U.S. court system seems less motivated than that in the EU to begin to dismantle intrusive surveillance techniques. Again, it appears that the only hope for private citizens is to devise workarounds of their own devising. Internet users have found ways to bypass government surveillance by using nontraditional search engines, such as Startpage, ixquick, and DuckDuckGo, billed as "the search engine that doesn't track you."

Duck founder Gabriel Weinberg said he named the search site after Duck, Duck, Goose, a circular-chase children's playground game. Since DuckDuckGo doesn't store previous searches, it does not and cannot present personalized search results. Weinberg said this frees users from the filter bubble—the fear that, as search results are increasingly personalized, they are less likely to be presented with information that challenges their existing ideas. He added that with no data storage, unlike other search engines, his site does not create a personal profile, thus becoming an advertising tool.

Weinberg said his opposition to widespread data gathering is philosophical. "I think of it as more privacy policy than general. I think [search engines] should be set up to be the minimal collection as needed, as opposed to the maximal collection possible. The other way to look at that is I think they should have a quid pro quo, which is 'you're giving up this particular piece of personal information and you're getting this benefit in return,' as opposed to the current status quo, which is 'we will collect anything we can and not tell you what the benefits are,' just say, in general, 'sure, you'll benefit from this.' I think that is the key difference. And you've seen some companies start to move to this direction, but very slowly." Weinberg's site apparently has struck a nerve with security-conscious Web surfers: DuckDuckGo handled more than a billion queries in 2013 alone.

In his classic book *Nineteen Eighty-Four,* British author George Orwell's protagonist Winston Smith was concerned about his television set, which received and transmitted simultaneously and recorded any sound above a low whisper. He never knew if he was being watched at any given moment. "You have to live—did live, from habit that became instinct—in the assumption that every sound you made was overheard, and, except in darkness, every movement scrutinized," wrote Orwell.

This picture of life in Orwell's fictional totalitarian state was published in 1948 yet seems eerily prescient of life in America today.

If the citizens of today do not begin to strenuously object to the ever-growing surveillance and intrusiveness of the police state and demand changes from their leaders, their children and grandchildren will grow up in an Orwellian authoritarian nightmare mostly unaware of true freedom and liberty.

CHAPTER 16

THE MILITARIZATION
OF POLICE

POLITICIANS REFUSE TO USE POLICE OR MILITARY FORCES TO FORtify and protect U.S. borders, but within the country's boundaries it seems that anything goes. Police forces are being armed with surplus war materials to tyrannize the civilian population, and even some school districts are accumulating military equipment for campus police.

There was a time in America when only criminals were fearful of the police. But that helpful public servant, that man in blue sworn to "serve and protect," has today been replaced by a black-clad, body-armored, and heavily armed enforcement officer who generally views ordinary public citizens as "perps." Increasingly, many young police officers and even entire law enforcement departments are arming and training as if they expect to be going to war with the American people.

This phenomenon has been encouraged at the federal level. Much of the militarization stems from the Defense Logistics Agency (DLA), the Pentagon's largest organization for the logistical support of U.S. military services as well as civilian law enforcement. The DLA is headquartered in Fort Belvoir, Virginia, and employs twenty-seven thousand people. The DLA, under the 1033 program begun in 1990, is charged with disposing of $28 billion worth of military equipment per year. Its Law Enforcement Support Office estimates that about eight thousand state and local police agencies have received billions in hardware since 1997. Much of this hardware became surplus as both the missions and materials of the military changed through the years. The pace of militarization increased significantly after the 9/11 attacks.

Frank Scafidi, director of public affairs for the National Insurance Crime Bureau, says, "The government has done a pretty good job since 9/11 of beating into our heads all of the nasty things that could befall us at any moment—from a terrorist threat. Cops like nothing more than grabbing onto some cool new (to them, at least) piece of hardware—especially when it was obtained at no cost. But then what? If the gear is in the inventory and an occasion to use it develops, you can bet that it will be used. And why not?"

According to Pentagon data, during the Obama years the military combat equipment received by police departments includes 432 MRAPs (Mine Resistant Ambush Protected) vehicles; 435 other armored vehicles; 44,900 night-vision goggles, sights, and lights; 533 planes and helicopters; and 93,763 machine guns. Police departments also have received nearly 200,000 ammunition magazines, thousands of pieces of camouflage and hundreds of silencers. And the federal government provides not only weapons and heavy equipment, but also a vast array of more mundane items, including office furniture, cleaning supplies, power tools, computers, and even disparate items such as an exercise bike, a treadmill, and refrigerators.

According to *New York Times* reporter Matt Apuzzo, "The equipment has been added to the armories of police departments that already look and act like military units. Police SWAT teams are now deployed tens of thousands of times each year, increasingly for routine jobs. Masked, heavily armed police officers in Louisiana raided a nightclub in 2006 as part of a liquor inspection. In Florida in 2010, officers in SWAT gear and with guns drawn carried out raids on barbershops that mostly led only to charges of 'barbering without a license.'"

Such militarization has even reached into some schools. According to an investigation by Houston, Texas, television station KHOU, ten Texas school districts have availed themselves of the free or low-cost military hardware from the Pentagon, including M16 rifles, M14 rifles, automatic pistols, tactical vests, and even military vehicles. Even though Spring Hills School District police chief Charles Brawner explained that such weapons would be available only to tactically trained officers in the event of an emergency, critics question the need for such heavy armament. "We've seen how even much-less-lethal devices like Tasers and pepper spray get used inappropriately and end up harming children," remarked Brennan Griffin of Texas Appleseed, a group that monitors campus police policies.

Citizens, outraged over such militarization, are pushing back. After hearing citizen concerns, the city council of Davis, California, a small college town near Sacramento, directed the police chief to get rid of a $700,000 armored car. Mayor Dan Wolk said, "It's the kind of thing that is used in Afghanistan and Iraq. Our community is the kind of community that is not going to take well to having this kind of vehicle. We are not a crime-ridden city. When it comes to help from Washington we, like most communities, have a long wish list. But a tank, or MRAP, or whatever you choose to call it, is not on that list."

Even some cops are concerned. As retired New York police detective John Baeza puts it, "A profession that I was once proud to serve in has become a militarized police state. Officers are quicker to draw their guns and use their tanks than to communicate with people to defuse a situation. They love to use their toys, and when they do, people die. The days of the peace officer are long gone, replaced by the militarized police warrior wearing uniforms making them indistinguishable from military personnel."

William Norman Grigg, a columnist for LewRockwell.com, notes that the 1033 program specifies that any law enforcement agency receiving military equipment must use it within a year or return it. "If a police department wants combat gear, it can get it, and once the department has that gear, it will use it. In fact, the feds will require that they invent a 'need' where none exists. This helps explain why routine police calls are now treated as potential military engagements," writes Grigg.

Some law enforcement agencies obtain military gear out of fear of being outgunned by criminals, while others simply want to emulate the militarized forces they see on TV and in films. Ominously, an 2014 investigation by the TV and digital network Fusion revealed that 184 state and local police departments had been suspended from the 1033 program because of missing weapons or failure to comply with guidelines. "We uncovered a pattern of missing M14 and M16 assault rifles across the country, as well as instances of missing .45-caliber pistols, shotguns and two cases of missing Humvee vehicles," write Daniel Rivero and Jorge Rivas.

Tim Lynch, director of the CATO Institute's Project on Criminal Justice, says the 1033 program is "obviously very sloppy, and it's another reason that Congress needs to revisit this promptly. We don't know where these weapons are going, whether they are really lost, or whether there is corruption involved." Lynch found the possibility that these military weapons were being sold on the black market "very unsettling." Investigations into missing weapons have been continuing in 2015.

Nowhere has the militarization of police been more blatantly on display than in the violence that erupted in August 2014 in the St. Louis suburb of Ferguson. Tensions there mounted following the police shooting of eighteen-year-old Michael Brown, and when demonstrators from the black community took to the streets to protest, they were met by heavily armed and armored police in full battle gear. Jake Tapper of CNN reported that there was no violence and looting until police began using tear gas and stun grenades. "Nobody is threatening anything," reported Tapper. "Nobody is doing anything. None of the stores here that I can see are being looted. There is no violence."

Tapper turned his camera to the unit of police and added, "These are armed police. With machine—not machine guns— semiautomatic rifles, with batons, with shields, many of them dressed for combat. Now, why they're doing this, I don't know. Because there is no threat going on here. None that merits this." Comparing the scene to military action in Afghanistan, Tapper reported, "There is nothing going on on this street right now that merits this scene out of Bagram. Nothing. So if people wonder why the people of Ferguson, Missouri, are so upset, this is part of the reason. What is this? This doesn't make any sense."

Frederick Reese, writing for *MintPressNews*, agreed that the massive police response did more to aggravate than to improve the situation. "The response—which included the targeting and arrest of journalists—is proof of the Elaborated Social Identity Model [ESIM], which suggests that a group of angry people can either be goaded into a riot or defused, depending on how they are treated by police. The impression of the police officer screaming at the crowd with his rifle raised—as has been seen on multiple occasions in Ferguson recently—is as much responsible for the rioting as the protestors themselves, according to ESIM. It is also, in part, a reflection of the notion that 'if the only tool you have is a hammer, you treat everything as a nail.'"

Lost in the mass media coverage of the unrest in Ferguson was a similar shooting death of twenty-year-old white man in Salt Lake City on August 11. This death also caused public demonstrations yet the police response was quite different.

Dillon Taylor, along with his brother and a cousin, were confronted by city police outside a convenience store. Ordered to stop, Taylor continued walking and was fatally shot in the chest after police claimed he reached into his waistband. The officer who shot him was "non-white," according to Salt Lake City police chief Chris Burbank, in response to concerns of racial profiling given that Dillon Taylor was Hispanic..

Jerrail Taylor said his brother was wearing earphones and did not realize what was happening and that Dillon was reaching for his cell phone. "He got confused, he went to pull up his pants to get on the ground, and they shot him," he told the media. Adding to the questions over Taylor death was the fact that police found no weapons at the scene and later released Taylor's brother and cousin without charges.

Chief Burbank also caused concern when he announced that the policeman who shot Taylor was wearing a body camera that recorded the entire affair, but he declined to make it public, saying, "It would be wholly inappropriate to take the most vital piece of evidence that we have and put it out to the public prior to the officer having some due process."

During the following week, demonstrators marched in Salt Lake City to protest what they felt was a wrongful shooting by police but the comparison between this event and those in Ferguson was stark. Unlike Ferguson, there was no rioting or looting and the police presence was kept minimal. "If we show up wearing riot gear it says throw rocks and boulders at us," explained Burbank. "I didn't send officers out wearing riot gear to talk to the protesters who were out in front of the building yesterday."

The difference in protests as well as the police response was not lost on some commentators. "While friends of Taylor and some others have protested the shooting, there have been no riots or violence, no radicals streaming in to agitate, no national media interest at all, and complete indifference from the A[ttorney] G[eneral] and POTUS [President of the United States], in stark contrast to their concern for the late strong-arm robber Michael Brown," wrote Thomas Lifson in the Internet publication *American Thinker*. "It is hard to escape the conclusion that some unarmed victims of police shootings are more important to the political and media leadership of America than others."

Two thousand and eight Constitutional Party presidential candidate Chuck Baldwin relates an experience that to him showed the change of attitude in the policeman of today. "True story," he writes, "here in Montana, a small town police officer, who is assigned to the traffic division, was asked to speak to a church group. Mostly, he gives out traffic citations for minor violations. As he began his remarks, he said, 'I am a cop; I work every day among the dregs of society.' Really? People who get parking tickets and speeding tickets are the 'dregs' of society? That, my friends, is the mark of an unfolding police-state mentality. And, remember, this is from the heart and lips of a professing Christian."

In an open letter to "my friends in law enforcement," Baldwin wrote, "As honest and honorable as most of you men and women of law enforcement are, it is time that you come to grips with the fact that the current system emanating from Washington, D.C., controlling the attitudes, training, and tactics of police agencies is practically a carbon copy of history's most notorious totalitarian regimes. And if the Nuremberg trials proved anything, they proved that 'I was just following orders' is never justification for ignoring the greater moral laws of God and Nature.

"My dad told me that the policeman is my friend. I would still like to believe that; but it behooves my friends in law enforcement to prove it to me by personally making up your minds to vehemently resist the current trend of militarizing your profession and of turning our once-free republic into a police state. After all, you want us to be your friends, too, right?"

The International Association of Chiefs of Police (IACP), in a 2012 report, claimed that police use of force is rare. The IACP said less than 2 percent of the forty million people who had contact with police in 2008 reported the use of force or the threatened use of force. "In large part, the public perception of police use of force is framed and influenced by the media depictions which present unrealistic and often outlandish representations of law enforcement and the policing profession," the group stated in its report.

Yet, even 2 percent of forty million is eight hundred thousand, and since 2010 the U.S. Justice Department has opened more than twenty investigations into police departments across the country. Fifteen departments accepted consent decrees, including ones in New Orleans, Seattle, and Portland. Several other departments reached out-of-court agreements, while four police agencies that rejected the DOJ findings found themselves facing lawsuits. A March 2015 Justice Department report found the Ferguson police had established a pattern of constitutional violations due to city officials pressing for more aggressive law enforcement to generate revenues. The investigation into the Ferguson Police Department was continuing in early 2015.

Critics have pointed out that the aggressiveness of police has substantially increased in recent years concurrent with the use of military hardware. Thoughtful people wonder why police need tanks and armored cars to catch speeders, shoplifters, or burglars.

A 2014 ACLU study entitled "War Comes Home: The Excessive Militarization of American Policing" concludes, "Across the country, heavily armed Special Weapons and Tactics (SWAT) teams are forcing their way into people's homes in the middle of the night, often deploying explosive devices such as flash-bang grenades to temporarily blind and deafen residents, simply to serve a search warrant on the suspicion that someone may be in possession of a small amount of drugs. Neighborhoods are not war zones, and our police officers should not be treating us like wartime enemies. However, the ACLU encountered this type of story over and over when studying the militarization of state and local law enforcement agencies."

This study substantiates the fact that police tactics nationwide are becoming "unnecessarily and dangerously militarized," mostly through federal programs providing military equipment to police agencies. It also found that the use of military equipment and tactics, which it called a "pervasive problem," was primarily being used against "communities of color."

"Historians looking back at this period in America's development will consider it to be profoundly odd that at the exact moment when violent crime hit a fifty-year low, the nation's police departments began to gear up as if the country were expecting invasion—and, on occasion, to behave as if one were under way," wrote Charles C. W. Cooke in National Review.com.

One example of the militarization of the nation's police forces, all now a national force under the control of the Department of Homeland Security, may be found in the small town of Neenah, Wisconsin. This quiet town of about twenty-five thousand has not recorded a homicide in more than five years, yet today is the proud owner of a thirty-ton MRAP combat vehicle. Neenah city councilman William Pollnow Jr. questioned the necessity for the MRAP but was told it was needed to protect police. "Who's going to be against that? You're against the police coming home safe at night? But you can always present a worst-case scenario. You can use that as a framework to get anything," he said. "Somebody has to be the first person to say 'Why are we doing this?'"

It is with notable irony that while U.S. troops in Iraq and Afghanistan early on were sent into combat with old-style jeeps and Humvees susceptible to destruction by improvised explosive devices (IEDs), local police forces are now being equipped with the new armored fighting vehicles during a time of relative peace in the country. In fact, while national leaders do everything they can to reduce the number of firearms in the hands of the citizenry, massive amounts of military hardware are being handed over to local police forces. Many wonder if this is simply a means of reusing old equipment in preparation for the purchase of newer models or if someone is preparing for major violence in America.

Some researchers find it more than coincidental that the militarization of the nation's police coincided with heightened public awareness of the globalists' designs and public debate concerning what President George H. W. Bush described as the "New World Order."

It should be further noted that the aggressiveness of police has followed training by Israeli antiterrorism experts, as will be detailed further on.

All this militarization began in the 1990s when Congress, sidestepping the 1878 Posse Comitatus Law that prohibits the U.S. military from policing the American public, authorized a military transfer program to aid local police against well-armed drug gangs. This delivery of military equipment continues today, although FBI statistics show crime has fallen to its lowest levels in a generation.

Police departments across the nation say they are simply preparing for worst-case scenarios by building up combat equipment. Captain Tiger Parsons of the Buchanan County Sheriff's Office in northwest Missouri, which now possesses a mine-resistant truck, says, "When you explain that you're preparing for something that may never happen, they get it." But not everyone agreed. Police Chief Ronald E. Teachman of South Bend, Indiana, declined the offer of a mine-resistant vehicle for his city. "I go to schools, but [instead of a war machine] I bring *Green Eggs and Ham.*"

It is true that with the possible exception of certain large cities, the vast majority of the nation's officers have not, and most probably will not, fire a weapon in a real-life situation. But they all have been conditioned by cop shows on TV to anticipate high-speed chases and gun battles as normal daily activities. No longer do police officers present themselves at a person's door and respectfully ask questions after showing proper identification. Today many citizens, whether guilty of some crime or not, are rudely awakened by busted-down doors and pointed guns in the hands of an armored SWAT team.

CHAPTER 17

THE RISE OF SWAT TEAMS

Since the early 1980s, the use of SWAT (Special Weapons and Tactics) teams has dramatically increased by more than 1,500 percent. A significant portion of this increase can be attributed to search and arrest warrants related to drug cases. And the nature of SWAT use, which has been encouraged by federal funding incentives, has changed as well.

According to a 2014 ACLU report, SWAT teams in the United States conduct around forty-five thousand raids each year, only 7 percent of which have anything to do with the hostage situations that were the original impetus for the creation of the teams. The ACLU reports that paramilitary operations are "happening in about 124 homes every day—or more likely every night—and four in five of those are performed in order that authorities might search homes, usually for drugs." These raids routinely involve armored personnel carriers, military equipment like battering rams, and flash-bang grenades.

The ACLU studied eight hundred deployments of SWAT teams among twenty local, state, and federal police agencies just in 2011–2012. They found only 7 percent were conducted for hostage, barricade, or active shooter situations. The study found most were drug searches, with just under 80 percent, or eight out of ten SWAT raids, intended simply to serve a search warrant, meaning they targeted someone only suspected of a crime.

More than 65 percent of the SWAT deployments studied, almost none with outside oversight, involved forcible entry, the use of a battering ram, explosive devices, or simply a boot. "And because these raids often involve forced entry into homes, often at night, they're actually *creating* [emphasis in the original] violence and confrontation where there was none before," opined Radley Balko, the *Washington Post* blogger and author of the book *Rise of the Warrior Cop: The Militarization of America's Police Forces.*

Although police routinely cite the presence of weapons to justify an armed and violent SWAT raid, the ACLU report observed that weapons were found in only 35 percent of the cases studied.

The ACLU said its report is incomplete, due to the fact that of the 225 law enforcement agencies asked for public records (out of more than 17,000 total U.S. agencies), 114 denied access to such records in whole or in part. "In short, we have police departments that are increasingly using violent, confrontational tactics to break into private homes for increasingly low-level crimes, and they seem to believe that the public has no right to know the specifics of when, how and why those tactics are being used," noted Balko.

Many SWAT teams are essentially accountable to no one. These independent SWAT groups claim that although they are funded by the taxpayers, they have incorporated and therefore, as private corporations, they are exempt from opening their records to the public.

For example, in Massachusetts, several SWAT teams operate under law enforcement councils, or LECs, funded by police agencies and overseen by an executive board, usually composed of police chiefs from member police departments. Critics point out that outsourcing SWAT team activities is an effective way for police departments to hide their military-style tactics and equipment from citizen scrutiny.

After citing a number of baseless drug raids and accidental deaths in the Bay State, Jessie Rossman, a staff attorney for the Massachusetts ACLU, said, "You can't have it both ways. The same government authority that allows them to carry weapons, make arrests, and break down the doors of Massachusetts residents during dangerous raids also makes them a government agency that is subject to the open records law."

These secretive SWAT enterprises have already caused a number of tragedies. In May 2014, a SWAT team in Habersham County, Georgia, in a botched drug raid, crashed through the door of a home and tossed a flash-bang grenade that landed in the crib of nineteen-month-old Bounkham "Bou Bou" Phonesavanh. Police claimed they had staked out the home for months based on erroneous information from an informant that someone living there was selling methamphetamine. According to the child's mother, "I heard my baby wailing and asked one of the officers to let me hold him. He screamed at me to sit down and shut up and blocked my view, so I couldn't see my son. I could see a singed crib. And I could see a pool of blood. The officers yelled at me to calm down and told me my son was fine, that he'd just lost a tooth. It was only hours later when they finally let us drive to the hospital that we found out Bou Bou was in the intensive burn unit and that he'd been placed into a medically induced coma."

Later in the year the family's attorney reported Phonesavanh has to date incurred an estimated $800,000 worth of expenses due to his injuries. Shortly after its personnel severely burned "Baby Bou Bou" with a flash-bang grenade, the Habersham County Sheriff's Department vowed to pay for the child's medical expenses. Later the family discovered through medical providers that the county had reneged on its promise. County officials stated that it would be "illegal" to pay.

Many felt it was bad enough that the incident happened but to cripple the family with nearly a million dollars in medical expenses was unconscionable and the refusal to take responsibility for a horrible mistake appeared criminally negligent, at the very least.

The raid on the home of Georgetown financial adviser Mark Witaschek one night in mid-2013 involved more than three dozen SWAT team members of the D.C. police who entered his home with guns drawn. Witaschek's fourteen-year-old daughter opened the door. Once inside, the SWAT team began pointing firearms at everyone in the home, including Witaschek and his girlfriend, Bonnie Harris. Witaschek said, "They used a battering ram to bash down the bathroom door and pull [his sixteen-year-old son] out of the shower, naked . . . The police put all the children together in a room, while we were handcuffed upstairs. I could hear them crying, not knowing what was happening." He added that the police search of his house caused an estimated $10,000 damage.

What caused this furor? Witaschek's estranged wife, in an attempt to obtain a restraining order against her husband, told a court clerk she had been threatened with a gun, a charge a judge later found was without merit. However, Washington, D.C., has the most stringent gun laws in the nation, which outlaw both weapons and ammunition if not registered. An avid hunter, Witaschek had stored his weapons with a sister in Arlington, Virginia. During the raid, police seized a holster, a brass casing, a box of slugs for use in a black-powder rifle (muzzle-loading muskets are not legally considered weapons), and one shotgun shell, which had misfired years earlier and was kept as a souvenir of a hunting trip. Although the shell casing and slugs could not be fired through a weapon, under the draconian D.C. laws, they still count as ammunition.

During a trial in March 2014, some of the evidence was removed from the record due to an insufficient warrant. By reducing the charges, officials were able to skirt a jury trial where they may have risked jury nullification. Witaschek would eventually be convicted of "attempted possession of unlawful ammunition" even though the judge never ruled on the shotgun shell, and the black-powder balls were for a weapon that is not considered illegal. A number of observers remarked that the prosecution had no clear understanding of either firearms or firearm terminology. "None of these people know anything about gun issues, including the judge," Witaschek charged. He was ordered to register as a convicted gun offender, which prompted him to tell the media, "I was found guilty of something that is not even illegal and forced to register for something that is not illegal. [Now] I run the risk of losing my job, my occupation, as a result of this conviction."

Some saw Witaschek's ordeal as an illustration of the misuse of SWAT raids and his conviction as a clear example of the end result of gun registration and police militarization. Many feel that when the feds desire to punish someone, they can always trump up any number of charges and have them arrested. Today, because of the militarization of law enforcement, they can also stage armed, SWAT-style raids of homes, businesses, and farms. "When the USDA has automatic weapons, night sights, and thirty-round magazines as part of a department of 'agriculture,' something has gone terribly wrong in America," observes *NaturalNews* editor Mike Adams.

In August 2014, tensions in Ferguson, Missouri, already were high after days of demonstrations following the fatal shooting of eighteen-year-old Michael Brown on August 9. There was nation-wide dismay over what many perceived as overreaction by local police, who descended on demonstrators with flak jackets, military equipment, and automatic weapons. The ensuing violence resulted in Missouri State Police taking control of the situation.

Only one Missouri police officer, who was not named at the time, was removed from duty during the Ferguson violence. This action came after the officer was videotaped pointing a weapon at bystand-ers at the demonstrations in Ferguson and telling a journalist, "I'm going to fucking kill you!" His removal was the result of an irate letter from the ACLU to Missouri Highway Patrol superintendent Corporal Ron Repogle stating, "This officer's conduct—from point-ing a weapon, to threatening to kill, to responding with profanity to a request for identity—was from start to finish wholly unacceptable. Such behavior serves to heighten, not reduce, tension."

Many would be shocked to learn that the often brutal and ag-gressive police tactics found in the news items of today may have originated in Israel, a nation surrounded by militant neighbors, which has long taken an aggressive stance toward its enemies. Israeli authorities have practiced constant crackdowns on Palestinian pro-testers and zero tolerance for terrorists.

According to Gordon Duff, senior editor of *Veterans Today,* the overreach of local police began during the George W. Bush admin-istration, when Michael Chertoff, who holds dual Israeli-American citizenship, became director of Homeland Security. Duff believes Chertoff mandated that American police forces be trained by Israeli groups in crowd control, counterterrorism, and intelligence gathering. As Duff notes, "Since that time, shootings of unarmed civilians have gone up 500 percent, attacks on legal political protests by police have become a scandal and huge stockpiles of ammunition and military heavy weaponry have been distributed to law enforcement groups in every region of America, both local and federally controlled.

Duff realized the extent of the problem with police killings when he entered the phrase "police shoot unarmed man" into Google, and got 1,610,000 responses. He said he found in almost every case that the officers had been trained in the Israeli martial art of Krav Maga, a fighting technique for use in violent encounters and combat situations that was originally developed by Israeli military personnel. Krav Maga has been encouraged by Homeland Security to deal with America's "fringe elements," including the poor, homeless people, mental patients, and veterans. The website of Krav Fit, a chain of schools that teach Krav Maga, claims that hundreds of U.S. law enforcement agencies have adopted this fighting skill.

During the 2014 violence in Ferguson that resulted after the fatal shooting of Michael Brown, many commentators compared the police response to Israeli aggressiveness toward Palestinian demonstrators. Not surprisingly, many police officials in the area, including St. Louis County police chief Timothy Fitch, received anti-terrorism training in Israel.

In October 2011, an exercise at the University of California at Berkeley saw joint training between American police forces and the Yamam, described by award-winning journalist Max Blumenthal as "an Israeli Border Police unit that claims to specialize in 'counterterror' operations but is better known for its extra-judicial assassinations of Palestinian militant leaders and long record of repressions and abuses in the occupied West Bank and Gaza Strip."

A month after these exercises, code-named "Urban Shield," the Alameda County Sheriff's Department was involved in a confrontation with the "Occupy" movement in downtown Oakland in which demonstrators were attacked with tear gas and rubber bullets, which left an Iraq War veteran in critical condition and dozens injured.

Blumenthal wrote in the Internet edition of the Arab newspaper *Al Akhbar* that since 9/11 there has been an "Israelification" of U.S. police forces, a circumstance not mentioned in the major media. "Having been schooled in Israeli tactics perfected during a sixty-three-year experience of controlling, dispossessing, and occupying an indigenous population, local police forces have adapted them to monitor Muslim and immigrant neighborhoods in U.S. cities," he noted. "Meanwhile, former Israeli military officers have been hired to spearhead security operations at American airports and suburban shopping malls, leading to a wave of disturbing incidents of racial profiling, intimidation, and FBI interrogations of innocent, unsuspecting people. The New York Police Department's disclosure that it deployed 'counter-terror' measures against Occupy protesters encamped in downtown Manhattan's Zuccotti Park is just the latest example of the so-called War on Terror creeping into everyday life. Revelations like these have raised serious questions about the extent to which Israeli-inspired tactics are being used."

Blumenthal said this intertwining of Israeli and American police forces has been documented through occasional news reports, but these reports typically highlight Israel's national security prowess without examining the problematic nature of working with a country accused of grave human rights abuses.

One example is the Georgia International Law Enforcement Exchange (GILEE), a police exchange program in which high-ranking Georgia police officers travel to Israel to learn counterterrorism tactics from the national police.

"Conversely, Israeli police officials travel to Atlanta every two years to learn Georgia's drug enforcement tactics," reported the *San Francisco Bay View* in 2011. "Through GILEE, the Israeli police adopt these tactics and employ them on Palestinian citizens of Israel and Palestinians residing in the occupied West Bank. While GILEE has relationships with several international police agencies, its relationship with the Israeli police is the most intimate and most troubling. Israel is one of the most brazen violators of human rights and international law in the world."

At a time when American police are being criticized for over-zealousness and aggressiveness, it is troubling for many to learn that they are receiving training from a nation that is being investigated by the International Criminal Court (ICC) for war crimes against Palestinians.

Adding to mounting concerns over the Israeli indoctrination of U.S. police were the 2014 statements of a former Israeli soldier. Eran Efrati, a twenty-eight-year-old sergeant in Israel's Nachal Division, served in the West Bank and was discharged in 2009. He then joined Breaking the Silence, an organization of Israeli veterans hoping to raise awareness of questionable activities in the West Bank.

Addressing the American public, Efrati warned, "If you don't care about Palestinians . . . you guys should know: you are next in line. The next one who will die from a tear gas canister into his chest will be in Zuccotti Park, will be in Denver, will be in Oakland, in San Francisco. It is happening here already. It is happening to different people, to people of color, to immigrants in this country, it is already happening. You guys are next in line. The next one who will die out of brutality of the police will be one of your sons or your daughters—in a protest. Because they [U.S. police and Israeli soldiers] are training together. Your police—training with our army. Our army is training them how to take care of the enemy . . . But when they come back, you are their enemy."

Many observers fear not only that this sort of official training may be negatively affecting the behavior of this nation's officers, but that certain aspects of American culture may also be to blame. Vicious films and video games may be partly responsible for hardening the minds of young people toward violence. Young men and women who play first-person shooter games and watch films filled with car chases, beatings, and shootings soon assume this is natural. These same conditioned youths grow up to be both police officers and gang members to whom violence is an accepted part of life.

Overly aggressive cops make for problematic situations, because even when cops are in the wrong, it is often difficult to discipline them. For example, in late 2013, Lafayette, Indiana, lieutenant Tom Davidson was videotaped knocking over the wheelchair of twenty-five-year-old Nicholas Kincade during an encounter with police. Lafayette police chief Patrick J. Flannelly, following an internal staff investigation, said Davidson's use of force was "outside our training and policy" and recommended Davidson be fired. But a civil service commission board voted three to two to merely have Davidson demoted and placed on probation following a thirty-day unpaid suspension.

CHAPTER 18

DEADLY FORCE

AN ADDED DIMENSION TO SUCH PROBLEMS IS THE INCREASING USE of private police forces. In 2012, Sharpstown, a community of 66,000 located southwest of Houston, joined more than seventy Texas communities in contracting with the private firm Strategic Executive And Logistical Security Solutions LLC (SEAL) to provide routine police protection.

Private police, termed by some as mercenaries for corporate America, are entrusted with all the powers of a government cop but not held to the same legal standards, according to Rutherford Institute President John W. Whitehead, adding, "[T]hese private police firms enjoy the trappings of government agencies—the weaponry, the arrest and shoot authority, even the ability to ticket and frisk—[but] they're often poorly trained, inadequately screened, poorly regulated and heavily armed. Now if that sounds a lot like public police officers, you wouldn't be far wrong."

Whitehead said the label of "private" is dubious at best. "Mind you, this is a far cry from a privatization of police. These are guns for hire, answerable to corporations who are already in bed with the government. They are extensions of the government without even the pretense of public accountability," he stated.

Of course, police misconduct is most worrisome when it involves lethal force.

The militarization of police has been accompanied by a new "us-versus-them" attitude toward the public. Police have not only become heavily armed but also quicker on the trigger. This is an issue that transcends issues of race.

And it's nothing new. According to *USA Today*, local police reported at least four hundred police killings to the FBI each year in the seven years ending in 2012. But despite the perception that a disproportionate number of blacks are being killed by white police, fatal police shootings appear to be an issue transcending racial divisions. Only ninety-six of the reported four hundred deaths were black citizens shot by white police. However, such reports did reveal that 18 percent of blacks killed during those years were under age twenty-one, compared to 8.7 percent of whites.

Nevertheless, *USA Today reporters* Kevin Johnson, Meghan Hoyer, and Brad Heath noted the database used for these statistics is considered flawed and largely incomplete. "The killings are self-reported by law enforcement and not all police departments participate, so the database undercounts the actual number of deaths. Plus, the numbers are not audited after they are submitted to the FBI and the statistics on 'justifiable' homicides have conflicted with independent measures of fatalities at the hands of police," they reported.

News reports collected over the past few years have indicated a marked increase in police misconduct and overreaction. But finding comprehensive data on police killings apparently is often difficult, if not impossible.

D. Brian Burghart, editor of the *Reno News & Review* spent two years attempting to assemble a national database of deadly police violence, but had little success. "Nowhere could I find out how many people died during interactions with police in the United States," wrote Burghart. "Try as I might, I just couldn't wrap my head around that idea. How was it that, in the twenty-first century, this data wasn't being tracked, compiled, and made available to the public? How could journalists know if police were killing too many people in their town if they didn't have a way to compare to other cities? Hell, how could citizens or police? How could cops possibly know 'best practices' for dealing with any fluid situation? They couldn't."

Burghart did accumulate some statistics, which show that in 2012, police killed nine people in Shelby County, Tennessee, and in 2013, killed eleven. "Who the hell knew Memphis police were killing men at more than double the rate the cops were killing people in Albuquerque?" he pondered. "The biggest thing I've taken away from this project is something I'll never be able to prove, but I'm convinced to my core: The lack of such a database is intentional," said Burghart. "No government—not the federal government, and not the thousands of municipalities that give their police forces license to use deadly force—wants you to know how many people it kills and why. It's the only conclusion that can be drawn from the evidence. What evidence? In attempting to collect this information, I was lied to and delayed by the FBI, even when I was only trying to find out the addresses of police departments to make public records requests. The government collects millions of bits of data annually about law enforcement in its Uniform Crime Report, but it doesn't collect information about the most consequential act a law enforcer can do."

Burghart's allegation is supported by Geoff Alpert, a criminologist at the University of South Carolina, who also points out that only about 750 of the 17,000 police agencies in the U.S. have contributed to the FBI's database. "There is no national database for this type of information, and that is so crazy. We've been trying for years, but nobody wanted to fund it and the [police] departments didn't want it. They were concerned with their image and liability. They don't want to bother with it," said Alpert.

The "bother" explanation rings hollow since all police department are bothered daily with trivial lawbreaking, such as jaywalking, slight speeding, and expired inspection stickers. It is more likely that departments, perhaps at the urging of federal authorities, do not want police violence publicized out of aversion to a public demand for stronger local control, not to mention the fear of liability issues.

The corporate mass media, accustomed to praising and supporting any police action, have largely overlooked the increasing use of deadly force by police. Next only to westerns, TV cop shows depicting brave and effective police officers have long been a stable of evening entertainment. The acceptance of police officers acting outside the law in order to bring justice to the bad guys, on TV and in movies, has not been lost on either the public or the cops.

But today there is a growing discontent with the number of deaths caused by newly militarized police agencies.

Some examples of unrestrained lethal force include the June 27, 2013, shooting of eighty-year-old Eugene Mallory, killed in his bed when Los Angeles County deputies used a no-knock warrant looking for methamphetamines. None were found.

In 2010, Douglas Zerby was finishing up watering his lawn, holding a hose nozzle, when neighbors, probably conditioned by the violence on TV, called police and reported him holding a gun. Long Beach police arrived and without warning fatally shot Zerby twelve times in the chest, arms, and legs. Zerby's family won a $6.5 million federal suit against the Long Beach Police Department, but his mother told the media, "The money does not bring my son back, which is all I really want."

In May 2014, the Purcellville, Virginia, police were called when seventeen-year-old Christian Alberto Sierra became depressed and unruly and cut himself with a knife. As he ran from his home, a policeman arrived and ordered him to stop. Then shots were fired and Sierra was fatally wounded. State police said the boy was shot after lunging at the officer, who was placed on administrative leave. The boy's distraught mother wondered why the policeman used such force on a teen they knew was in emotional distress, not committing a crime. "Why would you shoot a child that is suicidal? You are there to save him, not finish him off," she wondered.

Like Sierra's mother, many citizens are becoming alarmed over such indiscriminate shootings by increasingly militarized and combat-minded police. One of the most egregious and publicized instances was the shooting of Miriam Carey, the Connecticut mother who was shot while making a U-turn in front of the White House on October 3, 2013.

The corporate mass media initially styled the incident as a terrorist threat, then backpedaled, reporting that Carey was probably on drugs or suffering from mental problems. Well into 2014, at least one search engine still categorized Carey as "the 34 year old Connecticut woman who was shot to death by law enforcement officers on October 3rd, 2013, after ramming a White House checkpoint gate and striking 2 police officers, with a 1 year old child passenger." After none of this allegations proved true, the story was quickly dropped.

Carey family attorney Eric Sanders was finally able to obtain the woman's autopsy report some six months later. It revealed no drugs of any kind in the body of the victim. Apparently, her erratic driving behavior was caused by being shot in the back of the head by White House guards after making an abrupt U-turn when she realized she had mistakenly turned into the White House main gate. The guards shot at her as she was leaving, with no evidence that she was a threat. She was then shot again in her car with her child secured in the backseat. There was no word on any disciplinary action against the officers involved and no one in Congress appeared interested in investigating or holding any one responsible for Carey's death.

One video that went viral over the Internet was the 2013 execution-style shooting of thirty-seven-year-old Daniel Saenz while in custody. The El Paso bodybuilder was struggling with Officer Jose Flores in the El Paso County Jail while being transferred for medical treatment, when the Flores pulled his service pistol and fired one shot into Saenz. A spokesman for a Texas law enforcement organization said a nearby guard hit Flores's hand, causing him to shoot Saenz, whose hands were handcuffed behind him at the time. Despite a security video of the entire affair, a grand jury decided not to bring any charges against Flores, who was placed on leave.

In an opinion piece for Vice.com, Natasha Lennard voiced dismay that the Saenz shooting prompted no outrage on the part of the public. "It is merely my humble opinion that seeing a cop shoot an unarmed man dead should produce a collective rage so strong that the police can feel it, see it, and smell it," wrote Lennard. Such rage was exhibited in Ferguson a year later and was met with a militarized police response resulting in violence on both sides.

Columnist William Norman Grigg noted, "Every day, somewhere in this supposedly free country, some version of this script is played out: A police officer spies an individual committing a harmless but 'illegal' act, aggressively pursues the subject, inflicts physical violence on the victim, then escalates that violence to lethal or nearly lethal levels when the victim doesn't immediately submit to the state-licensed aggression."

Even conservatives have expressed concern over the number of deaths at the hands of police. Economist Paul Craig Roberts, a former assistant secretary of the treasury for economic policy under President Reagan and a former associate editor of the *Wall Street Journal,* commented in the wake of the 2014 police killing in Ferguson, "The gratuitous violence employed by police is no more justified than the gratuitous violence employed by the Israeli military toward Palestinians. 'Law and order conservatives' confuse police accountability with the coddling of criminals and terrorists. They are unable to comprehend that unaccountable police are a greater threat to them than are criminals without badges."

There is usually no public anger over misconduct by increasingly aggressive police. But such anger did assert itself in Albuquerque, New Mexico, a city of 555,000 that witnessed twenty-six deaths by police in just four years. As in other places, the deaths were followed by some controversy and a debate over how much force is necessary for a police department to do its job. But events in Albuquerque took a dramatic turn in May 2014, following the murder of James Boyd, a mentally ill homeless man who was fatally shot after forty officers descended on him as he attempted to camp out in the Sandia foothills. Albuquerque citizens, irate over the number of police shootings, stormed a city council meeting and even tried to make a citizens' arrest on the police chief. The meeting was canceled with all but two council members fleeing the room.

The council canceled another meeting and set prohibitions on signs and protests, but demonstrators continued to voice their discontent. "When are they going to quit killing people and start taking them into custody?" asked Ken Ellis, there to protest the killing of his son, a twenty-five-year-old Iraq War veteran suffering from post-traumatic stress disorder who was gunned down in 2010. "They have to address this issue. They can't sweep it under the rug anymore."

In January 2015, Bernalillo County district attorney Kari Brandenburg, responding to the public outcry over police killings, announced she would lodge murder charges against two of the Albuquerque police officers involved in Boyd's death. Brandenburg said unlike recent high-profile cases in Ferguson and New York City, a public trial would allow the public to hear all the witnesses and view the evidence.

In April 2014, the U.S. Justice Department issued a forty-six-page report detailing a pattern of excessive force on the part of Albuquerque police, including a policy of shooting at moving vehicles to disable them, a policy allowing officers to use personal weapons instead of standard-issue firearms. According to this report, "Officers see the guns as status symbols. APD personnel we interviewed indicated that this fondness for powerful weapons illustrates the aggressive culture."

In response to the report and in an apparent effort to stop leaks, Albuquerque police chief Gorden Eden sent a text message to all department employees instructing them to stop meeting with DOJ investigators. A spokesperson for the police explained the chief's text message was to ensure that all communications to the DOJ came from official negotiations, including those of the chief.

In the wake of the DOJ report, the APD banned the use of personal guns and the practice of shooting to disable moving cars along with other reforms, policy changes that should be emulated by other departments.

In May, irate citizens in Hearne, Texas, marched on police headquarters following the death of ninety-three-year-old Pearlie Golden. Upset over being denied a driver's license, she reportedly brandished a gun when police arrived at her home.

The citizen demonstration prompted the city council in Hearne to vote unanimously for the firing of Police Officer Stephen Stem over the fatal shooting. About a year previously, Stem had been placed on leave after fatally shooting a suspect during the investigation of an unruly crowd in an apartment parking lot.

Public anger over police killings also arose in Los Angeles and Las Vegas. The militarization of police forces has only added fuel to the fire of anger and suspicion over what many see as police heavy-handedness.

Sometimes fatal police action does not even require the use of bullets. In mid-2014, an Illinois lawsuit was filed in federal court against six police officers accused of killing a ninety-five-year-old World War II vet with beanbag bullets.

John Wrana Jr. had resisted leaving his assisted living center to go to a hospital for treatment of a suspected urinary tract infection. Police were called when Wrana refused to leave his room. According to the complaint filed by Wrana's stepdaughter, police decided to take the man by force. One officer fired five rounds of beanbag bullets from a twelve-gauge shotgun from a distance of about six to eight feet. A medical examiner ruled that the man's death was due to blunt-force trauma to his abdomen as a result of the shots.

The lawsuit stated that documents from the beanbag manufacturer noted that the beanbags can travel at approximately 190 miles per hour and that "shots to the head, neck, thorax, heart or spine can result in fatal injury." Rob Kall, publisher of *OpEd News,* said, "When a brutal fascist authoritarian law is put out there, by legislation or executive decree, the ripples it produces can be massive. The president may order the killing of two or three Americans and hundreds or thousands of others. He may authorize agencies to secretly detain a person or a few people indefinitely. But then the slippery slope leads to people being detained without the president even knowing—for reasons of 'plausible deniability.' When the president can do it, every petty, two-bit judge, sheriff, town cop, magistrate and courthouse clerk starts thinking it's okay to push the limits."

Even honest citizens, many with police or military backgrounds, find themselves on the receiving end of increasingly authoritarian government tactics. Captain Nicolas Aquino, while attending the Naval Post Graduate School in Monterey, California, was confronted outside his home by Sheriff's Deputy Ivan Rodriguez. Apparently, the deputy had been summoned by a neighbor of Aquino's who called to report a suspected burglar. Even after explaining he lived there and showing his military ID and offering his utility bill, Aquino was placed in a choke hold and handcuffed.

After studying Aquino's wallet, the deputy realized that the captain was indeed the legal resident. But Aquino related, "The officer did not apologize. He pulls me over to the side of the driveway and he does basic victim blaming, and he says it was my fault for not knowing my neighbors. He then states that he had wanted to Tase me if he had a Taser, and he would have shot and killed me if he had drawn his weapon, and he would have been fully justified in killing me."

In May 2014, a Tulsa policeman was accused of firing his gun at a car containing two teenagers who were making out late at night in an empty school parking lot. When the officer approached their parked car, the teens tried to drive off but were stopped when the officer shot out their tires. The officer claimed he felt threatened by their attempt to leave and fired his weapon. It was not explained how firing a gun in a neighborhood was somehow less threatening than scared teens trying to drive off to avoid getting in trouble.

It is apparent from the abundance of such stories that police departments should require more extensive training—not in weapons and tactics but in how to deal fairly with the public and to be more aware of the need for good public relations. They should be reminded that they work for the citizens. They are not on the streets to act as parents or prison guards.

But most importantly, officers must avoid the shoot-first-ask-questions-later mentality and understand that lethal force must be the last resort in any conflict.

CHAPTER 19

WRONGFUL ARRESTS

ONE ALL-TOO-COMMON RESPONSE TO BOTH UNWARRANTED SUR-veillance and the growing police state goes as follows: "If you are not doing anything wrong, you have nothing to worry about."

In today's world, with its ever-expanding list of laws and regulations, coupled with the fear generated by the federal government and corporate media in the aftermath of 9/11, it is possible for a person to get in serious trouble even when doing nothing wrong. Consider these real cases from recent years.

Niakea Williams's son has Asperger's, and she was called to his elementary school because he had had a behavioral episode. In her haste she failed to properly sign in to the school, and despite the fact that the staff knew her, the whole school was placed on lockdown and Mrs. Williams was arrested and escorted out in handcuffs.

Or consider the case of Greg Snider, of Houston, Texas, who was pulled over on a freeway and surrounded by ten cop cars after a drug sting gone wrong. Snider had stopped in a downtown parking lot to make a business call, when he was approached by a homeless man who asked for spare change. Snider gave the man seventy-five cents and drove off. Moments later, his car was stopped, and the police were shouting, "We saw you downtown. We saw what you did." "I was like, 'Are you kidding me? I gave a homeless man seventy-five cents,' " said Snider, who was held for a half hour before police realized they had mistaken his act of charity for a drug deal.

In 2012, Paul Valin was about fifteen miles from his Des Moines home when he found a backpack that contained what he believed may have been meth-making equipment. He reported his find to the authorities and, as a consequence, his home was listed on the DEA's National Clandestine Laboratory Register as a meth lab. The Justice Department's website admits that the "Department does not accept responsibility or liability for damages of any kind resulting from reliance on an entry."

Also in 2012, New York police staged a midnight raid on the Bronx home of Gerald Bryan, punching through walls and tearing out light fixtures looking for drugs, even though they had no warrant to do so. Bryan was arrested and $4,800 in cash was seized. One year later, the case against him was dropped. But when he tried to get his money back, he found that under asset forfeiture laws it had been deposited into the pension fund of the NYPD.

Bryan's case is merely one example of civil forfeiture, the act by which a municipality can seize money during an arrest. Civil forfeiture, which has become increasingly common as cash-strapped cities look for additional revenue streams, has been upheld as constitutional by the U.S. Supreme Court, which ruled that a person can lose their property even if someone else used the property to commit drug crimes without one's knowledge. A former head of the forfeiture unit for the Bronx DA's office admitted that in about 85 percent of civil forfeiture cases, the property owner is never charged with a crime.

By early 2015, the forfeiture practice had drawn such public criticism that U.S. attorney general Eric Holder announced he was restricting the federal government's role in a civil asset protection program in which federal agencies could take possession of assets seized by local law enforcement with the exception of firearms or other materials related to public safety. Previously, such assets could be handed over to the feds, who would share proceeds with the local entities. Holder said since all states now have their own asset forfeiture laws, it is no longer necessary for the Justice Department to hold seized assets.

Random drug testing is another way in which police use public fear of illegal drugs to terrorize law-abiding citizens. In 2013, Jameson Hospital and Lawrence County Children and Youth Services agreed to pay $143,500 to settle a lawsuit filed by Elizabeth Mort, whose infant daughter was taken away for five days because of a false-positive drug test apparently caused by consumption of a poppy seed bagel. In 2014, Rachael Devore sued another hospital, Magee-Womens at the University of Pittsburgh Medical Center, saying a false-positive drug test apparently caused by poppy seeds in farmers-market bread resulted in an Allegheny County Children Youth and Families investigation of her family.

According to a story by Radley Balko in the *Washington Post*, "No one would argue that a new mother with a drug habit presents all sorts of problems for both mother and child. But even if these tests were 100 percent accurate, treating both patients for addiction seems like a far more humane policy than yanking a newborn from his mother's arms—or sending the mother to prison."

Jordan Wiser, a Jefferson, Ohio, high school student, was another casualty of police hysteria fueled by warrantless intrusions into private citizens' lives. Wiser was held in jail for thirteen days and finally allowed to post a $50,000 bond after officials of the technical school he was attending used the student handbook as a warrant to search his vehicle, where they found two airsoft guns and a folding-blade pocketknife.

Another high school student, in Clarksville, Tennessee, David Duren-Sanner, who had never been in trouble and was carrying a three-point grade average, was suspended for ten days and sent to an alternative school after police found a fishing knife in the car he was driving during a lockdown. The senior student had given authorities permission to search his car believing he had "nothing to hide." The car belonged to his father, a commercial fisherman. Duren-Sanner's grandmother, Peggy Duren, said she tried explaining that the knife didn't belong to her grandson, but school officials wouldn't listen. "Unfortunately [the vice principal] said that's the way it is now: Guilty until proven innocent. It's part of this zero tolerance policy," she said, adding that should the boy's punishment be upheld, Duren-

Sanner will not be able to attend prom, his JROTC ball, or walk at graduation.

Seventy-year-old Washington State resident Darien Rossen filed a lawsuit against the Idaho State Police after Trooper Justin Klitch stopped Rossen's car, searched it, and accused Rossen of using marijuana. It seems Rossen's car carried license plates from Colorado, a state that has legalized pot. Klitch claimed Rossen's eyes "appeared glassy," but no drugs were found. These are not isolated incidents. They are part of an ominous pattern of inflexible laws combined with police overreaction. And there are many other examples of people being punished for doing nothing wrong.

Sometimes the alleged crime they've committed isn't even on the books in the United States. One unlucky man found out you can go to jail for violating an obscure law of another country. Abner "Abbie" Schoenwetter had imported fish from Honduras for twelve years, packaging them in plastic bags, all inspected by both U.S. Customs and the FDA. But then the National Marine Fishery Service decided that Schoenwetter had violated a Honduran law requiring cardboard packaging rather than plastic. Despite the fact that he had no criminal record and that the attorney general of Honduras stated that the cardboard regulation was not applicable in his case, Schoenwetter was convicted of violating another country's law and sentenced to six years in prison.

This case was not a fluke, according to Brian W. Walsh, senior legal research fellow in the Heritage Foundation's Center for Legal and Judicial Studies. Walsh notes aspiring inventor Krister Evertson spent almost two years in federal prison because federal prosecutors and EPA officials, aided by questionable federal laws, decided that storing materials in three-eighths-inch-thick stainless-steel drums was the equivalent of disposing of them without a permit.

Three-time Indianapolis 500 race-car champion Bobby Unser was arrested and convicted of driving a snowmobile on a federally protected wildlife area even though Unser and a companion said they had mistakenly entered the area lost in a blinding blizzard for two days and almost died.

In November 2013, Jason Dewing was ticketed by an upstate

New York officer for driving while talking on his cell phone. When his case came to trial in March 2014, he was able to demonstrate, using his phone records, that he had not, in fact, been talking on the phone, but was instead using an electronic cigarette. So the judge then charged and convicted Dewing of violating New York traffic law 1225-d, driving while using a portable electronic device, even though an e-cigarette cannot make a call or send text or data.

Jeff Sherwood of the *Knoxville Business Examiner* commented: "It is a sad day in America when one driver can pass an officer smoking a tobacco cigarette and not warrant a second glance, but when another driver passes an officer 'vaping' an electronic cigarette, he finds himself in front of a judge trying to explain the law."

So what are we to do? Former defense attorney James Duane, now a professor at Regent Law School, advises people that it is never a good idea to talk to police. He explained that even the Congressional Research Service cannot count the total number of federal crimes on the books as they are scattered throughout more than fifty titles of U.S. Code filling some twenty-seven thousand pages. He noted that the American Bar Association (ABA) estimates that additional administrative regulations issued by Congress total nearly ten thousand.

Duane's advice was supported by Supreme Court Associate Justice Stephen Breyer, who has stated, "The complexity of modern federal criminal law, codified in several thousand sections of the United States Code and the virtually infinite variety of factual circumstances that might trigger an investigation into a possible violation of the law, make it difficult for anyone to know, in advance, just when a particular set of statements might later appear (to a prosecutor) to be relevant to some such investigation."

And these are only *federal* laws.

Entire books have been written about squirrelly laws still on the books in many states. For instance, you can't play dominoes on Sunday in Alabama, nor can you view a moose from an airplane in Alaska. It's a violation for women to wear pants in Tucson, and it's illegal to eat an orange while sitting in a bathtub in the state of California. And there's plenty more where these came from.

Often these bizarre laws are not enforced, but sometimes they are, to disastrous effect. In September 2014, the West Virginia Natural Resources Police made eleven arrests and seized 190 pounds of ginseng, a plant highly valued in Asia as a medicine. Police said the ginseng haul was worth an estimated $180,000. That same month, twenty-five persons were arrested for illegally harvesting ginseng in southern Indiana. Who knew harvesting ginseng was illegal?

It turns out that growing and using ginseng on one's own property is legal, but harvesting the plant out of season or on public land and all national parks is an offense.

And, of course, there are always the old selectively enforced standbys that can get you arrested. Loitering, failing to show proper identification, and, certainly not least, trying to argue with an officer or, worse yet, trying to pull away from their grip, which means you are evading police or assaulting the officer, which can send you to jail quickly, if not get you beaten or shot.

In the police state that is expanding within the American death culture, one does not even have to be home to receive a warrantless search by authorities. A little-known police tactic is the increasing use of "sneak and peek" warrants that allow police to covertly enter private homes, perform searches, seize property, and then leave quietly without notifying the homeowner.

Sneak-and-peek warrants, an extension of "no-knock" raids, are obtained by having a judge authorize the police to secretly break into private property without first announcing themselves or presenting the subject of the search with a signed warrant. The search is usually conducted when the suspect is not at home and, in many cases, accomplished surreptitiously, even to disguising the break-in to look like a burglary.

"The entire premise [of sneak and peek] encourages government agents to adopt the tactics of criminals in order to gain access to property: breaking and entering, sneaking around, stealing, and risking a surprise confrontation with an unsuspecting civilian," concluded analysts at the PolicestateUSA website.

One tactic offered to protect families from both would-be robbers and no-knock police would be to simply build a three-foot "waiting area" extension to the front door so that whoever might break down the first door, alerting the residents, would only face a second one.

One shocking way to give overzealous police second thoughts about no-knock raids and the sanctity of a person's home was illustrated in early 2014 when a Texas grand jury declined to impose a capital murder indictment against a man for killing a sheriff's deputy who entered his home in a no-knock raid searching for some pot plants. This decision was tantamount to jury nullification (when a jury decides on the justness of a law as well as the letter of the law) and reflected the public attitude in Texas that a man's home is his castle and should remain inviolate except in the most extreme cases.

Twenty-year-old Goedrich Magee, who shot and killed Burleson County sergeant Adam Sowders during a late 2013 raid, said when officers broke into his rural home, he thought he was being robbed and acted to protect his pregnant girlfriend and children. Magee was charged with possession of marijuana after deputies recovered a handful of plants and seedlings.

"Once again, there has been little attention to the increasing no-knock warrants that have grown in tandem with the militarization of our police forces," wrote constitutional law scholar Jonathan Turley. "The result is not just a chilling effect for citizens but increasing mistaken shootings. In this case, an officer is dead and the prosecutors wanted to send away a father for life—for a raid to secure a few marijuana plants."

Yet despite numerous false arrests and deadly encounters between police and citizens, there has been little outcry over the state of policing in the USA. Rutherford Institute founder John W. Whitehead remarks, "What is most striking about the American police state is not the mega-corporations running amok in the halls of Congress, the militarized police crashing through doors and shooting unarmed citizens, or the invasive surveillance regime which has come to dominate every aspect of our lives.

"No, what has been most disconcerting about the emergence of the American police state is the extent to which the citizenry appears content to passively wait for someone else to solve our nation's many problems. Unless Americans are prepared to engage in militant nonviolent resistance in the spirit of Martin Luther King Jr. and Gandhi, true reform, if any, will be a long time coming."

After recounting the growth of government spying, militarized police forces, the shooting of unarmed civilians, the erosion of personal property rights, the loss of personal integrity through strip searches, drone surveillance, and the criminalization of children's behavior, Whitehead concludes, "To put it bluntly, we are living in an electronic concentration camp. Through a series of imperceptible steps, we have willingly allowed ourselves to become enmeshed in a system that knows the most intimate details of our lives, analyzes them, and treats us accordingly."

Following incidents of police deaths in New York and Missouri, the U.S. Justice Department announced it was widening its involvement with local police in an attempt to curb the growing abuse of police power.

"The arrogance and heavy-handed tactics of . . . rogue police officers is largely attributed to the influence of the U.S. government's Department of Homeland Security. Now this same U.S. government is going to 'widen' its involvement with local police agencies in order to curb police violence? Get real! You know what's happening: the U.S. Justice Department is widening its involvement in local police agencies in order to further its parochial political agendas and to further promote political correctness," warned Chuck Baldwin.

The growing power of the DHS, which now has jurisdiction over all the U.S. police forces, coupled with overreaching enforcement of innumerable laws and regulations, worry those who view this behemoth as a homegrown Gestapo, the secret police of Nazi Germany. The head of the Gestapo, Heinrich Himmler, controlled all of Germany's police forces.

A Baptist preacher, Baldwin said what is needed are constitutional sheriffs who will serve as "the people's vanguard against both the overreach of the federal government (which spawns so much police abuse, both local and federal) and against rogue officers among the police agencies of his jurisdiction." He also recommended partisan-free prosecutors in the local court systems, because many prosecutors seem motivated by partisan politics rather than "liberty and justice for all." "And we also need jurors who are truly blind to their prejudices and propensities and are willing to acquit or convict on the basis of proven fact alone. When a policeman steps over the line, he or she must be held as accountable as any ordinary citizen," he added. This would seem especially true in cases involving private police forces.

It's clear there needs to be a buffer between the citizenry and police. Some have suggested a civilian review board (CRB) might alleviate the situation.

A CRB is composed of community residents with no connection to the police who meet to review public complaints about police misconduct and offer recommendations for improvements to police chiefs and lawmakers. CRBs may be an effective way to assuage fear and suspicion in a community. Members should be drawn from a broad cross section of the population, with an effort to balance race, religion, and education. But care must be taken to avoid authorities packing a CRB with police cronies and sycophants. Unfortunately, past experience with review boards has shown that they are easily co-opted by political authorities and end up supporting internal police investigations.

For many years, most police agencies have opposed the notion of civilian oversight, but many thoughtful officers today are warming to the idea. Randy Rider, writing in the police publication Officer. com, notes that CRBs have been used successfully in Georgia, particularly in the supervision of child fatalities, child abuse, and domestic violence.

"Initially there was great resistance to these boards. However, they are now accepted institutions and have created a sense of security ensuring that these cases are investigated thoroughly. In some cases their oversight has led to new procedures, laws, and revealed evidence not discovered prior to the review," argued Rider. Some activists and even some review boards have been surprised to find that local-department-internal-affairs investigations have often imposed harsher punishments than those sought by a CRB.

"Any police department in this country ought to be proud to have inspected any internal matter that they have investigated. It would display good will and instill a sense of confidence to the public," argued Rider. "After all, who are we here to protect? Are we above being held accountable? We are held to a higher standard than the average citizen. We have to be. We are in the public eye. How can we go to court and take another person's freedom if we are not?"

CHANGING THE GAME

CITIZENS MUST ASSUME GREATER RESPONSIBILITY FOR THEIR OWN actions, learn to work with one another for justice and true equality, and send a stern message that they will not tolerate a police state.

It is now obvious that U.S. police departments must begin an effort toward demilitarization. If the nation is to avoid a total police state, law enforcement should return to proven and kinder methods of crime prevention and control, especially as statistics show that crime is decreasing across the country. Walking patrols in some inner-city neighborhoods may bring police into closer contact with the citizens even if this means less traffic fines. And individuals must take it upon themselves to monitor the actions of police. This can vary from judicious use of their cell-phone camera to lodging complaints when the situation warrants it.

But primarily, public policy must focus on the basic causes of community unrest and crime: lack of jobs, educational opportunities, and hope. The poor should be encouraged to feed themselves nutritious food from small gardens rather than be fined and discouraged by overbearing local ordinances.

Local governments, the churches, and the media should make community service a "cool" activity instead of a punishment. They should encourage feeding the poor and homeless. Public relations campaigns could raise awareness on matters of community relevance. Local governments should encourage the start of programs such as the Works Projects Administration (WPA) of the Great Depression era in which the unemployed were put to work on public projects such as roads, dams, and waterways. Applicants for unemployment compensation could be put to work in jobs provided by local government. They could not be fired from such public jobs but would be required to register and show up for work each workday. Putting people in jobs would not only restore dignity to those out of work and reduce the welfare rolls, but might actually produce much-needed public improvements. Taxpayers would be getting something for their money.

As Benjamin Franklin once noted, "Only a virtuous people are capable of freedom. As nations become more corrupt and vicious, they have more need of masters."

CHAPTER 20

FINANCE CAPITALISM

CAPITALISM LIES AT THE HEART OF AMERICAN SOCIETY. THE legacy of abuse of this system is death and misery. Though it's an imperfect measure, we can begin to see the impact of finance capitalism in the spike in suicides in areas affected by the recent financial collapse. Since the beginning of the worldwide financial recession in 2008, there have been more than ten thousand suicides in North America and Europe, according to the *British Journal of Psychiatry.* Researchers said this nearly 10 percent rise in deaths was due to lost jobs, lowered home prices, stock losses, and dwindling pensions.

Furthermore, since 2002 the number of workers collecting Social Security Disability Insurance (SSDI) checks grew 48 percent, half of whom claimed disability due to mental or mood disorders to include back pain. Dead and disabled Americans are among the most obvious reminders of the rise of giant multinational corporations and banks whose globalist leadership is advancing an agenda of increasingly tight control and depopulation—and all for enormous profit.

The globalist agenda has been revealed for years but almost never discussed in the corporate-controlled mass media. As far back as 1966, Carroll Quigley, a Georgetown University professor of history and a mentor to former president Bill Clinton, himself a member of an international anglophile network he said operated in secrecy,

wrote in his book *Tragedy and Hope,* "The powers of financial capitalism had a far-reaching aim—nothing less than to create a world
system of financial control in private hands able to dominate the
political system of each country and the economy of the world as a
whole. This system was to be controlled in a feudalist fashion by the
central banks of the world acting in concert, by secret agreements
arrived at in frequent private meetings and conferences."

Quigley explained that central banks, including the Fed, would
dominate governments by their ability to control treasury loans, to
manipulate foreign exchanges, to influence the level of economic activity in the country, and to "influence cooperative politicians by
subsequent economic rewards in the business world."

Global banking interests, centered in the Bank for International
Settlements (BIS) and the International Monetary Fund (IMF),
along with little-known agencies, can be found at the center of efforts to destabilize sovereign nations across the world, converting citizens into debt slaves to the globalist banking system.

Because of a growing list of broadcast and Internet talk-show
hosts and causes such as the "Occupy" movement, more and more
Americans have become aware of the idea of a New World Order
composed of wealthy globalists and their banking interests. This
concept was considered merely a conspiracy theory only a few short
years ago, but today the snickering has subsided as the effects of this
agenda become more obvious every day.

HIDDEN TOOLS OF FINANCE

GLOBALISTS, THAT ONE PERCENT OF THE POPULATION COMPOSED OF
super-rich individuals who own or control the earth's major corporations and resources, are highly adept at using little-known financial
mechanisms in order to gain control over the world economy. One
such hidden tool is the Exchange Stabilization Fund (ESF) of the
United States Treasury. This little-known agency has morphed into
a gigantic money-moving operation with absolutely no oversight by
the U.S. government, the states, or the American people. Yet the ESF

is supported by taxpayer dollars and uses its funds to carry out covert operations both inside and outside the country.

The ESF was created and originally financed by the Gold Reserve Act of 1934, then considered one of the most important bills to come before Congress since the Civil War because it shifted U.S. financial polices from Federal Reserve banks to the treasury, and hence, the ESF. Its purpose, according to an official description, was "to contribute to exchange rate stability and counter disorderly conditions in the foreign exchange market." The act authorized the secretary of the treasury to deal exclusively in gold, foreign exchange, securities, and instruments of credit subject to the approval of the president. The ESF is the mechanism by which short-term loans are made to foreign governments. As of December 2014, the ESF held more than $95 billion in assets.

When ownership of gold was outlawed for Americans in 1933, the ESF transferred gold out of the country to foreigners in exchange for dollars, in turn draining our gold reserves for years to come. Today, when people blame the Federal Reserve for stealing America's gold reserves, they should know that it was actually the ESF, the agency never scrutinized or questioned by Congress, that drained the Federal Reserve many years ago.

In 1978, under agreements with the IMF, Congress amended the Gold Reserve Act to allow the ESF to provide short-term credit to foreign governments and monetary authorities. These ESF "bridge loans" are financed through currency swaps, which means dollars held by the ESF are made available to a country through its central bank in exchange for the same value in that country's currency. The ESF also administers special drawing rights (SDRs), assets created by the IMF then loaned to countries requiring help in financing balance-of-payment deficits. SDRs are permanent resources of the ESF, whose operations are conducted through the Federal Reserve Bank of New York. The New York Fed acts as an intermediary between the ESF and those foreign governments seeking short-term financing.

The independent nature of the ESF has alarmed many observers. According to some members of the House Committee on Coinage,

Weights, and Measures who reviewed the gold reserve law during its preparation, "This [law] in fact, means that the Secretary of the Treasury shall be under no obligation to comply with general laws of the United States in the handling of this fund . . . We believe that [this] places autocratic and dictatorial power in the hands of one man directly over the control of the value of money and credit and indirectly over prices . . . We believe that this is too great a power to place in the hands of any one man. We believe that it is contrary to every true principle of American Government."

In 1940, treasury officials began calling the ESF its "secret weapon," as millions of dollars were loaned to China to resist the Japanese invasion and to Argentina to defend against Nazi influence. Some isolationist members of Congress were unhappy with these expenditures of American tax money and were doubly unhappy to learn that the ESF was under no obligation to follow U.S. laws.

The ESF has continued to hold tremendous sway over global financial policy and has been described as the largest financial agency in the world. After World War II, ESF chief Harry Dexter White helped design the world's new monetary system, including its best-known creations—the International Monetary Fund (IMF) and the World Bank.

White concurrently served on many government committees including one that founded the Office of Strategic Services (OSS), forerunner of the CIA. Today, the ESF has copycat agencies in many states and itself has been accused of being a slush fund for CIA "black operations." There are even accusations of laundering drug money. The fund operates outside of legislative oversight and public scrutiny and, while technically legal, outside of the intentions of the U.S. Constitution and Bill of Rights.

According to Lawrence Houston, the first general counsel of the CIA, "The heart and soul of covert operations is . . . the provisions of un-vouchered funds, and the inviolability of such funds from outside inspection." This is accomplished through hidden mechanisms such as the ESF.

In 2008, when the Federal Reserve borrowed $90.3 billion to bail out the bankrupt American International Group (AIG), the funds came through the ESF, where some staffers worried that Congress might object if the ESF's role was discovered.

Former presidential candidate Dr. Ron Paul has called the ESF a "slush fund" hidden within the Federal Reserve System. "The initial funding of the program came from taking the gold in from the people and then revaluing up so they had some money. So it's off-budget budgeting. Congress doesn't keep appropriating money. They [the ESF] can earn enough interest buying Treasury bills . . . But they can interfere in the market.," explained Paul.

The hidden nature of the ESF has caused the media to refer merely to "monetary authorities" when it means the ESF and allows the Fed to take the blame for any mistakes.

Eric de Carbonnel, a prolific blogger at MarketSkeptics.com, has produced a video series revealing an immense amount of information on the ESF. He concluded that the agency "controls the New York Fed, runs the CIA's black budget, and is the architect of the world's monetary system (IMF, World Bank, etc.). ESF financing (through the OSS and then the CIA) built up the worldwide propaganda network which has so badly distorted history today (including erasing awareness of its existence from popular consciousness). It has been directly involved in virtually every major U.S. fraud/scandal since its creation in 1934: the London gold pool, the Kennedy assassinations, Iran-Contra, CIA drug trafficking, HIV, and worse."

The Financial Stability Board (FSB) is another example of the global elite's hidden centralized control over the economy. The FSB was created in 2009 when President Obama signed an agreement at a London meeting of the G20.

Marilyn M. Barnewall, described by *Forbes* as the "dean of American private banking," noted, "It seems the world's bankers have executed a bloodless coup and now represent all of the people in the world . . . President Obama agreed at the G20 meeting in London to create an international board with authority to intervene in U.S. corporations by dictating executive compensation and approving or disapproving business management decisions . . . Under the new Financial Stability Board, the United States has only one vote. In other words, the group will be largely controlled by European central bankers. My guess is, they will represent themselves, not you and not me and certainly not America."

The FSB was merely an outgrowth of the Financial Stability Committee of the Bank for International Settlements (BIS), which was controlled by the Nazis until after World War II. The FSB will oversee the Federal Reserve System, the Securities and Exchange Commission, and other federal agencies, thus effectively placing U.S. economic policy in the hands of international bankers.

This handful of bankers is behind the flow of money necessary for the mergers and acquisitions of other banks. Once there were hundreds of banks in America, owned by individuals and local families. But due to government regulations put into place during the Reagan-Bush years, these banks either faded away or consolidated. In 1990, there were thirty-seven major banks in the U.S. By 2009, buyouts, mergers, and bankruptcies had reduced this number to four: Citigroup, JPMorgan Chase, Bank of America, and Wells Fargo. Ominously, in June 2012, the giant global rating agency Moody's downgraded the ratings of Bank of America, Goldman Sachs, and JPMorgan, citing concerns for the stability of the world's financial system. The International Monetary Fund's 2014 Global Financial Stability Report, noted that efforts to stabilize the world's financial systems were far from complete and conditions "far from normal." The report stated that the probability of more taxpayer bailouts of "too important to fail" banks remains high.

A LONGSTANDING PLAN

PUBLIC APPREHENSION OVER THE PRACTICES OF LARGE BANKS IS nothing new. Citizens have long been concerned by the state of the banking industry. In 1922, former New York City mayor John F. Hylan warned, "The real menace of our Republic is the invisible government, which like a giant octopus sprawls its slimy legs over our cities, states and nation . . . a small group of powerful banking houses generally referred to as the international bankers. The little coterie of powerful international bankers virtually run[s] the United States government for their own selfish purposes. They practically control both parties, write political platforms, make cat's-paws of party leaders, use the leading men of private organizations, and resort to every device to place in nomination for high public office only such candidates as will be amenable to the dictates of corrupt big business."

Even worse, Hylan explains, they control the flow of information to the general public. "These international bankers and Rockefeller–Standard Oil interests control the majority of the newspapers and magazines in this country. They use the columns of these papers to club into submission or drive out of office public officials who refuse to do the bidding of the powerful corrupt cliques which compose the invisible government. It operates under cover of a self-created screen [and] seizes our executive officers, legislative bodies, schools, courts, newspapers and every agency created for the public protection."

The long-standing plan to financially subjugate citizens of the U.S. is no secret. The average citizen understands it better than the corporation-controlled mass media. The plan was articulated in 1924 by Montagu Norman, governor of the Bank of England. Addressing the American Bankers Association in New York City, Norman explained, "Capital must protect itself in every possible way, both by combination and legislation. Debts must be collected, mortgages foreclosed as rapidly as possible. When, through process of law, the common people lose their homes, they will become more docile and more easily governed through the strong arm of the government applied by a central power of wealth under leading financiers.

"These truths are well known among our principal men, who are now engaged in forming an imperialism to govern the world. By dividing the voter through the political party system, we can get them to expend their energies in fighting for questions of no importance. It is thus, by discrete action, we can secure for ourselves that which has been so well planned and so successfully accomplished."

This consolidation of wealth has changed the United States from a democratic republic under the rule of law with ordered liberty guided by the Constitution and Bill of Rights into a fascist state at the forefront of an effort to establish a New World Order ruled by the banking interests and controlled by multinational corporations.

American publisher and free-speech advocate Larry Flynt in a blog for the *Huffington Post* described the situation thusly, "The American government—which we once called our government—has been taken over by Wall Street, the mega-corporations and the super-rich. They are the ones who decide our fate. It is this group of powerful elites, the people President Franklin D. Roosevelt called 'economic royalists,' who choose our elected officials—indeed, our very form of government. Both Democrats and Republicans dance to the tune of their corporate masters. In America, corporations do not control the government. In America, corporations *are* the government [emphasis in the original]."

The late outspoken comedian George Carlin offered this unfunny view of the globalists: "They don't want well-informed, well-educated people capable of critical thinking . . . They want OBEDIENT WORKERS [emphasis in the original]. People who are just smart enough to run the machines and do the paperwork, and just dumb enough to passably accept all these increasingly shittier jobs with the lower pay, the longer hours, the reduced benefits, the end of overtime, and the vanishing pension that disappears the minute you go to collect it."

The financial crisis in the U.S. today, with its unequal income distribution, disproportionate executive pay, continuing housing and derivative bubbles, coupled with NSA surveillance, the Patriot Act, intrusive TSA manhandling, Homeland Security, ever-more-restrictive gun laws, militarized police forces, torture, and the classification of government critics as "potential terrorists," is looking more like the country portrayed in *The Hunger Games* films than the democratic republic handed down by the forefathers.

This situation did not come about overnight. Wealthy elites have attempted to control North America since the colonization of the continent began, and these efforts began to gain traction with the rise of the Robber Barons in the late nineteenth century. Efforts to curb the excesses of unfettered capitalism after the turn of the twentieth century have resulted in the centralization of power in the federal government, which has been given greater authority to regulate industry even as the creation of collective bargaining attempted to give more power to workers.

Yet rather than leveling the playing field of the owner-worker relationship, these efforts have only bolstered the few who hold real power on each side. Union power has declined sharply even as labor leaders hobnob with business owners in country clubs and business organizations.

In politics, money talks. In fact, it screams. Any politician knows that the surest way to winning an election is to buy huge amounts of time on television. TV networks and stations want political advertising paid up front and airtime is costly. This system almost assures that political office will go to whichever candidate spends the most on TV ads, not to whoever might be best qualified.

And the court system does not appear to be the slightest bit concerned by this arrangement. The Supreme Court recently ruled that corporations can be considered the same as individuals in that they may contribute unlimited amounts of money to the political candidate of their choice. This means large corporations can gain political support by simply opening up their pocketbook. In reality, the ability to buy political influence has widened the gap between the wealthy elite and the working poor.

The globalist elite are like parasites. They feed off the population all the while soothing us, even as they plot our extermination, as will be detailed later.

These are the same elitists who engineered the rise of the Bolsheviks in 1917 Russia but then grew fearful as the Bolsheviks turned communist and began urging class warfare worldwide. These same globalists next created national socialism in pre–World War II Germany in an attempt to confine socialism within national boundaries.

Hitler, a German army intelligence agent, was an underling in this global ruling elite's long-term ethnic extermination program, although he initially stated he was fighting international finance and loan capital. The globalists turned against Hitler only after he began building his Third Reich with Reichsmarks, interest-free money not borrowed from international bankers. After the war, thousands of unreconstructed Nazis were brought to America through such programs as Operation Paperclip. The Nazis who survived the war and their descendants have steadily built a Fourth Reich, expanding from the military-industrial complex into corporate leadership. Along with Nazi technology, they brought Nazi philosophy, helping to explain why this elite continues to rule many nations today.

Businessman and author J. R. Dunn, a consulting editor of the website Americanthinker.com, details the shift that has taken place in the U.S. "What does this transformation look like overall? It involves drastically cut energy sources, seriously limited forms of transportation, nineteenth-century (or even more primitive) health care, less food, expensive light sources, seriously truncated job opportunities, and higher education too expensive for any but the elite. In other words, Americans are to become poor, cold, hungry, and stupid, while sitting in the dark. That's the progressive plan for the wealthiest nation in the long human record," he wrote.

Dunn goes on to address the rationales being used to impose control over the public. "The reasons behind this don't matter; they are myriad, contradictory, and unconvincing—global warming, fairness, equality, sharing the wealth. They change from week to week at convenience. Because the only true reason is power. The elite does it because they can."

The placement of Nazi scientists in the American system after World War II has been well documented. The globalists who supported the Nazis merely moved them to the USA under projects such as Paperclip and implanted both technology and political philosophy within the military-industrial complex. For decades, they have sought to create a world government controlled by a scientific dictatorship and guided by the principles of eugenics—survival of the fittest.

WHO ARE "THEY"?

ONE ALWAYS HEARS OF "THEM." YET IT IS SIMPLISTIC TO BLAME THE amorphous "them" for everything that is wrong with contemporary society. "They" are greedy capitalists. "They" are behind the banking excesses, the military adventurism, and the social engineering. "They" control the news media. "They" want us dead. "They" want to reduce the world's population. "They" are trying to rule the world through world government.

But who exactly are "they"?

They are the global corporate masters who have guided modern America to a culture of death—from deadly drugs, food, water, and air to violent entertainment and blood sports. The culture of death created by greedy globalists has pervaded every aspect of American life.

To some conspiracy researchers, these globalists are collectively referred to as the Bilderbergers, a group of powerful men and women—many of them European royalty—who meet in seclusion each year reportedly to discuss the issues of the day. The group is called the Bilderbergers because its existence was first discovered by the public in 1954 after a meeting at the Bilderberg Hotel in Oosterbeek, Holland. In the past, both government and media officials declined to even mention the Bilderberg meetings, writing off concern over their gatherings as a conspiracy theory. But in recent years that has all changed. The alternative media, and even some mainstream news sources, have begun covering the annual meetings.

The official explanation for Bilderberg gatherings is that they are just friendly get-togethers of prominent business and government leaders and their discussions are kept private to encourage an open exchange of ideas. But just consider the cries of "restraint of trade," "monopoly," and "price fixing" that would erupt should, say, the owners of the National Football League franchises gather behind closed doors, surrounded by armed guards, and refuse to publicly say what they discussed.

A list of reported attendees from the secretive Bilderberg meeting held in Copenhagen in May 2014 can be found in the appendix. These names will not mean much to the reader because most of these men and women are rarely mentioned in the corporate mass media. Yet they collectively represent a concentration of wealth and power unequaled in the modern world.

These are a substantial number of "they" who once a year meet to deliberate the fate of national economies and, hence, entire populations. Many of them also believe in the mandate of eugenics, the practice of improving the human race to include reducing the population.

Know that we do not have the names of every attendee. Only those who authorize the release of their names get mentioned in the public media. Daniel Estulin, author of *The True Story of the Bilderberg Group*, wrote that the group's membership and meeting participants have represented a "who's who" of the world power elite with familiar names like David Rockefeller, Henry Kissinger, Bill and Hillary Clinton, Gordon Brown, Angela Merkel, Alan Greenspan, Ben Bernanke, Larry Summers, Tim Geithner, Lloyd Blankfein, George Soros, Donald Rumsfeld, Rupert Murdoch, other heads of state, influential senators, congressmen, and parliamentarians, Pentagon and NATO brass, members of European royalty, selected media figures, and invited others. Such invitees have included President Obama along with many of his top officials.

Estulin said that also represented at Bilderberg meetings are leading figures from the Council on Foreign Relations (CFR), IMF, World Bank, the Trilateral Commission, EU, and powerful central bankers from the Federal Reserve, the European Central Bank (ECB), and the Bank of England.

David Rockefeller, the head of the Rockefeller family financial empire, is believed to have been a leading Bilderberg attendee for years. Other wealthy elite members merely send representatives.

An official with Switzerland's Credit Suisse bank has estimated the entire net wealth of the world at more than $440 trillion, which is expected to rise in the coming years. By some estimates, the Rothschild banking dynasty reportedly controls up to $300 trillion in assets. In the US, the Rockefeller family is estimated by some to be worth about $100 trillion. If the estimates are accurate, these families collectively own or control virtually all the world's major banks, multinational corporations, conglomerates, and oil companies. They also own massive real estate holdings such as castles, palaces, stately mansions, luxury hotels, racetracks, casinos, exotic holiday resorts, along with large tracts of farmland and uncut forest in various countries, which explains the lack of certainty as to their wealth.

Mike O'Sullivan, a chief investment officer at Credit Suisse, told CNBC that the top richest one percent of the world's population owns 46 percent of global assets. And it is all centered in the world's largest banks.

A 2011 study by scientists at the Swiss Federal Institute of Technology in Zurich found that only about two dozen multinational banks—named were Barclays Bank, JPMorgan Chase, and Goldman Sachs—controlled upward of forty-three thousand transnational corporations.

George Sugihara, a complex systems expert and adviser to Deutsche Bank, remarked, "It's disconcerting to see how connected things really are."

Such a concentration of power, cemented by greed, wealth, and class loyalty, has led to policies perhaps not in the best interests of all humankind. It therefore comes as no surprise that it is within the global elite where the instigators of plans to reduce the human population must be found.

CHAPTER 21

DEATH OF THE SPECIES

Iᴛ's ɴᴏᴛ ᴏɴʟʏ ᴛᴏxɪᴄ ᴇɴᴠɪʀᴏɴᴍᴇɴᴛᴀʟ ꜰᴀᴄᴛᴏʀs, sᴛʀᴀɴɢᴇ ᴅɪsᴀʙɪʟ-ities, diseases, and authoritarian policies that are leading to the demise of the human race. For several years now, in many parts of the world, scientists and fertility doctors have found that men's sperm count and motility (sperm movement) are declining, making it more difficult for couples to have babies. One 2012 study in France estimated that for thirty-five-year-old men, all else remaining equal, sperm concentration dropped from 73.6 million per milliliter (ml) in January 1989 to 49.9 million/ml by December 2005. A fertility clinic study of seven thousand men in Aberdeen, Scotland, led by Dr. Siladitya Bhattacharya found that the average sperm count of those men with a normal sperm concentration (more than 20 million sperm per milliliter) in the group fell from nearly 87 million to just over 62 million over fourteen years, a 29 percent drop.

An earlier Danish study found sperm counts had dropped by one percent every year since 1938. Organizations such as the Center for Reproductive Epidemiology at the University of Rochester Medical Center have announced similar findings. Center director Dr. Shanna Swan, acknowledges that while such studies are far from definitive, the center's data place yearly sperm declines at 1.5 percent in the U.S. and 3 percent in Europe and Australia.

Oddly, although some researchers consider the continuing decline in sperm count a "crisis," no recent studies have been published.

University of Edinburgh professor Richard Sharpe comments, "'In the UK this issue has never been viewed as any sort of health priority, perhaps because of doubts as to whether 'falling sperm counts' was real. Now there can be little doubt that it is real, so it is a time for action. Doing nothing will ensure that couple fertility and average family size will decline below even its present low level and place ever greater strains on society."

Some see declining sperm counts as the result of radiation poisoning from Fukushima and Chernobyl, while others believe it is the result of harmful chemicals in the food and water supply. Some blame lower sperm counts on bisphenol A (BPA), an additive in plastics found in many household products or pesticides, as scientists have noticed that men from rural areas where farming pesticides are common have lower sperm counts than men from urban areas. Others cite sexually transmitted infections, stress, obesity, and even watching television as the culprits.

It would appear that the truth behind lowering sperm counts can be attributed to a combination of these environmental factors, along with the adulteration of the human food supply with toxic additives and chemicals, as previously described. All of this has been fostered by the aforementioned giant multinational corporations. These corporations are run by able administrators under the orders of shadowy owners, whose names are largely unknown to the public, but many may be found on the membership list of the Bilderbergers.

If these current trends continue, humanity's days on earth may be numbered. By recognizing the disastrous end awaiting the population and taking action now, the American public can begin to move toward solutions and remedies that can bring a more peaceful and prosperous future. Many of these problems begin with proper education.

THE DEATH OF REAL EDUCATION

IN A CULTURE DEDICATED TO LIFE, AFFORDABLE HIGHER EDUCATION should be offered to all those who wish to attend a college or university. There was a time in America when a young person from a family of limited means could work his or her way through college. But today such opportunity is often unavailable. According to the *Washington Post*, the average cost of higher education rose by 307 percent between 1989 and 2010, while the average income of workers rose a mere 70 percent.

Today, it is impossible for a working student to make enough for a college education. Back in 1979, an academic credit at Michigan State University cost $24.50. A student making the then minimum wage of $2.90 per hour could pay for an hour's credit with one day's work. Adjusted for inflation, that would be is $79.23 in today dollars. One credit hour today costs $428.75. Today, a student earning 2014's minimum wage of $7.25 an hour would need to work sixty hours to earn enough to pay for one credit hour.

What has led to such a drastic shift in the economics of higher education? Some have pointed to the large amounts of money giant corporations donate to colleges and universities. What might first appear as philanthropy, many see as merely instilling corporate views in the learning process.

A coalition of academics and activists have endorsed "A National Call: Save Civilian Public Education." Its website explains, "Over the last several decades, the Pentagon, conservative forces, and corporations have been systematically working to expand their presence in the K-12 learning environment and in public universities. The combined impact of the military, conservative think tanks and foundations and of corporatization of our public educational systems has eroded the basic democratic concept of civilian public education. It is a trend that, if allowed to continue, will weaken the primacy of civilian rule and, ultimately, our country's commitment to democratic ideals."

This group points to such programs as ROTC (Reserve Officers' Training Corps), the closed-circuit-TV Channel One beamed into eight thousand schools, corporate contracts for providing brand-name food and soft drinks, and the proliferation of private charter and "cyber" schools, arguing that these destroy the traditional objective of American public education. "The cumulative effect is the creation of institutions that cultivate a simplistic ideology that merges consumerism with subservience," they said.

And indeed, despite national efforts such as "No Child Left Behind," "Common Core," and programs like "school choice," and "recovery school districts," American public education has continued to decline for five decades.

Tom Allon, owner of Manhattan Media and a Republican candidate for mayor of New York City, writes, "First of all, our children are not being stimulated from an early age and many lose interest in learning by the time they are in elementary school. We think that the 'one size fits all' public education system, an industrial model designed in the mid-1900s, should work in this post-information and digital age. This is clearly wrong and we must now design curricula that set every child's mind 'on fire,' even if it means using digital technology much more in the classroom and incorporating online learning as well as animation and vocational training, for those who are not traditional academic learners."

Pointing to a study by the National Council for Accreditation of Teacher Education, Allon argues that teacher preparation today is woefully inadequate. "Training teachers is not a one-week series of seminars before their first days in the classroom. It's not a theoretical class in one of our educational graduate programs. It must represent at least three, if not four, years of vigorous apprenticeship as a student teacher before entering the classroom as a lead teacher and then a five-to-ten-year series of mentoring programs that are conducted by 'master teachers' or 'mentors,' two new tiers of teaching that I would recommend to remedy our teacher training and retaining crisis (50 percent of American teachers leave the profession in their first five years)." Allon especially points to one "intangible thing necessary to lift our country out of our downward spiral: R-E-S-P-E-C-T [emphasis in the original] for teachers and the teaching profession."

"The path to education victory is not as simple as A-B-C. But it's also not as hard as the Pythagorian theorem. It just takes a paradigm shift for our elected leaders to stop searching for scapegoats and start acting like real superheroes," wrote Allon. "Our kids—this generation and the next one—can't wait any longer. We need the fierce urgency of now to stop the educational insanity which plagues our society. But first we must put our teachers and students first—ahead of politicians and the testing industry."

Allon's call for respect was echoed by Dennis Van Roekel, president of the National Educators Association (NEA) in a speech before the 2012 NEA national meeting. Roekel noted, "We all know there are plenty of people who are eager to offer advice—or worse, try to impose their ideas on our profession—bloggers, columnists, elected officials, and self-proclaimed reformers, they are constantly weighing in about public education . . . they love to talk about and blame teachers. As if this disjointed and underfunded system is somehow the fault of those who teach and the people who work in those schools. But the real problems are the profiteers and mega-rich Wall Street folks who created an economic crisis that has our country and the world reeling. And the solution isn't to attack educators, it's to give respect. That's what will attract talented young people to become teachers and education support professionals and college professors."

Van Roekel says the reason we must support education is simple: "Public education makes America strong. Studying history and civics helps students become good citizens. Part of a democratic republic. Public education is a vehicle to teach American values and ideals, values like a just society, equal opportunity, and democracy. And in a nation where equal opportunity is one of our most deeply held values, education is a key that opens the door to economic opportunity, for people from all backgrounds."

Legal researcher Shayna A. Pitre, writing on *The Blog* said America could learn a few things from other countries whose students rate higher in education tests. "Only when the United States does this, and learns the right lessons from these countries' practices, will the era of education reform truly arrive," she said.

Pitre describes successful tactics employed by foreign teachers, tactics that lead to quality education. Teachers more highly trained in colleges that demand high grades, they train and complete student teaching before landing a job, they receive pay commensurate with other professions, they teach more critical thinking skills such as problem solving rather than rote memorization, and they use non-computerized international standard testing that requires students to work out complex problems.

Some suggestions to bring American education back its earlier success include: improving training for teachers; getting the best teachers to mentor the others; teaching both teachers and students to be critical thinkers and problem solvers; allowing students to learn at their own pace and in their own way as opposed to one teacher telling thirty-five kids to do the same thing at the same time; getting parents more involved in the education process; and addressing the issue of bunching together too many kids from impoverished backgrounds in inadequate schools.

America's Founding Fathers' writings were based on the concept of the worth and power of the individual. Individualism was once viewed as an admirable trait. It conjured up pictures of the rugged cowboy, honest lawman, and brave soldiers.

In the midnineteenth century, a great civil war was fought over states' rights, a term that has today become unfashionable and linked to bigotry and provincialism. Yet politicians still extol the virtues of an individual's rights. This dichotomy begs the question of how a person can have personal individual rights but not states' rights?

In today's world of political correctness, one can talk of the "individual" within certain limited contexts, noted Jon Rappoport, author of *Power Outside the Matrix.* "You can say 'power,' if you're talking about nuclear plants, or if you're accusing someone of a crime, but if you put 'individual' and 'power' together and attribute a positive quality to the combination, you're way, way outside the consensus. You're crazy."

Audiences today still see individual power extolled as a virtue in movies, television, video games, comics, and graphic novels. The entertainment industry presents a wide variety of cops, secret agents, spacemen, and superheroes and heroines who succeed on their individualism and wits.

"But when it comes to 'real' life, power stops at the front door and no one answers the bell," said Rappoport. "Suddenly, the hero, the person with power, is anathema . . . So he adjusts. He waits. He wonders. He settles for less, far less. He learns how the game is played. He stifles his hopes. He shrinks. He forgets. He develops 'problems' and tries to solve them within an impossibly narrow context. He redefines success and victory down to meet limited expectations. He strives for the normal and the average. For his efforts, he receives tidbits, like a dog looking up at his master. If that isn't mind control, nothing is."

Today, the globalists, through control of government, the education system, and the mass media, have advanced the idea of collectivism, subordinating the individual to the "greater good of the group." Young people are being conditioned to blindly follow instruction and learn by rote instead of honing the skills of independent thinking and deductive reasoning.

This must change. Liberty and an effective democracy demand citizens who can think critically and for themselves.

CHAPTER 22

DEATH OF THE MASS MEDIA

THE CORPORATE MASS MEDIA IS ANOTHER AREA THAT MUST BE RE-formed if we are to fight against the wishes of the global elite. Interestingly, there is evidence that the mass media behemoth is in decline.

In the 1960s, only three TV networks—ABC, CBS, and NBC—dominated the broadcast audience, creating a centrist consensus of life in the United States. During the weekend of the Kennedy assassination, the entire nation watched in shock as every major media outlet preempted normal programming and stories to deal with the tragedy.

The concentration of media was a double-edged sword to the corporate elite. On the one hand, official pronouncements could swiftly reach nearly the entire nation, while on the other, contradictory and antiestablishment messages could do the same if they broke into the major media. *Life* magazine exemplified this dichotomy by publishing evidence of conspiracy in the assassination even while supporting the government's lone-assassin theory.

Slowly, the rise of cable TV, along with the buyout of homegrown newspapers by large corporations, altered the media landscape. On September 11, 2001, subscribers could turn from the coverage of the 9/11 attacks to watch the Disney Channel or ESPN.

Since that time, listenership has become divided between broadcast, cable, satellite, and the Internet, and advertising revenues for the mass media have continued to decline, prompting some commentators to claim that TV is dead. It is true that where once a single episode of the western series *Gunsmoke* could capture more than 40 percent of the TV audience, today's most popular shows are lucky to garner 10 percent of the audience.

As far back as 1993, novelist Michael Crichton predicted that the major established media, which he termed "Mediasaurus," would become as extinct as the dinosaurs within a decade. Comparing the American corporate media to a used car, Crichton argued that it was of "very poor quality," and that "its information is not reliable, it has too much chrome and glitz, its doors rattle, it breaks down almost immediately, and it's sold without warranty. It's flashy but it's basically junk."

Critics today likewise view television news as quick and cheap programming that is repetitive, simplistic, and insulting. Cable-TV news is viewed as predominately unqualified talking heads, and newspaper reporting as mostly rewritten press releases full of unnamed sources.

Many see news stories today as no more than opinion pieces that reflect the zealotry and intolerance of advocates. Clay Shirky, a professor of new media at New York University and author of *Cognitive Surplus: Creativity and Generosity in a Connected Age*, noted, "Years ago, it wasn't necessarily news that people wanted to watch when they got home. They just wanted to watch TV, and the news was what was on. Once they were given the option of ESPN, viewers couldn't change channels fast enough. This removed the population of politically uncommitted viewers from the news audience, leaving only the partisans."

Some suggest that in view of the mass media's lack of objective and reflective studies of current events, Americans today live in an age of conformity much more confining than the 1950s.

Crichton was not the first commentator to point out the nefarious vapidness of the mass media. In 1967, Marshall McLuhan wrote *The*

Medium Is the Massage: An Inventory of Effects, detailing how the media controls content and how content is received by the individual. One glaring example of the corporate mass media's twisting of words can be found in his famous adage "the medium is the message." However, in both his book's title and in his conclusion, McLuhan stated that "the medium is the *massage.*" But today, the corporate media usually employs the word "message" rather than "massage," no doubt because media moguls do not want the average citizen to consider the idea that messages are being "massaged" before being brought to the public.

"As it turns out, the traditional television business is far stickier than people thought, and audience behavior is not changing as rapidly as people thought it might," said analyst Richard Greenfield of BTIG Research. "Yes, television viewing went down in 2012 for the first time, but people are still watching five hours a day. YouTube is growing, but people are watching eight minutes a day. They are where cable was in 1980." But in Greenfield's estimation, it will not take thirty years for the Internet and YouTube to surpass broadcast and cable television in viewership.

This is happening already. According to estimates by Wall Street media analysts Craig Moffett and Michael Nathanson, in the third quarter of 2013, cable companies lost 687,000 subscribers. "Viewers are abandoning their TV sets to watch on new devices and through new distribution channels," explained Shirky. "From 2011 to 2012, the number of videos streamed on tablets and smartphones rose 300 percent, with digital outlets like YouTube, Hulu, Netflix and Amazon capturing both new users and more time spent."

While predictions of the death of mass media might be premature, there does seem to be some truth in them, particularly in regard to the print media. "As we pass his prediction's fifteen-year anniversary, I've got to declare advantage Crichton," admitted Jack Shafer, editor-at-large for Slate.com in 2008. "Rot afflicts the newspaper industry, which is shedding staff, circulation, and revenues. It's gotten so bad in newspaperville that some people want Google to buy the *Times* and run it as a charity! Evening news viewership continues to evaporate, and while the mass media aren't going extinct tomorrow,

Crichton's original observations about the media future now ring more true than false. Ask any journalist."

While the U.S. was once a nation with a great variety of newspapers and periodicals, today virtually everything a person sees or hears is coming from one of only five multinational corporations—the Walt Disney Company, News Corporation, Time-Warner, and Viacom (which now includes CBS) and the German publishing giant Bertelsmann. These five giants not only control the newspapers but for most of them also radio and television networks, movie studios, magazines, cable and satellite outlets, music companies, and even billboards.

A study by Project Censored, a nonprofit media research group managed through the School of Social Sciences at Sonoma State University, revealed the largest media companies are actually interconnected by common owners and board members.

Within ten major media corporations, there were 118 individuals who sat on 288 different national and international corporate boards. The study also documented media directors who had served as former senators or representatives, revealing a "revolving door" relationship between corporate media and U.S. government officials.

Concentration of media ownership has resulted in progressively fewer individuals or organizations controlling increasing shares of the mass media, As more and more media companies fall victim to the transnational corporations through buyouts and takeovers, a media oiligarchy has been created that dominates the industry.

The late C. Edwin Baker, professor of law and communication at the University of Pennsylvania Law School, in his book *Media Concentration and Democracy: Why Ownership Matters,* questioned the support of deregulation and hypercommercialism demonstrated by current media ownership. Baker argued that dispersal of media ownership could result in more owners who would reasonably pursue socially valuable journalistic or creative objectives rather than a socially dysfunctional focus on the bottom line.

Mass media monopoly can mean programming representing only the agenda of its globalist ownership, undue loyalty to both government and corporate advertisers, and censorship of free discourse in the public interest.

Concentration of media has led to fights over deregulation. Proponents of deregulation argue that the removal of government rules will allow commercial exploitation and thus increase profits, encourage more diverse ownership, and aid developing nations in acquiring their own media companies.

Opponents say deregulation will only result in a more dangerous concentration of ownership by globalist corporations, reducing the diversity of information and opinions as well as the overall quality of programming.

Though the business of television may appear healthy from the outside, it's clear that a decline in the industry is ongoing. Even Fox News, long considered a success story of cable news, suffered the lowest audience numbers in more than a decade in 2014, according to an article on the Politico website. And a huge proportion of viewers of Fox News (and other cable networks) are senior citizens. Younger viewers, the audience of the future, simply are not there.

Hadas Gold of Politico shows just how dire the demographics are. "Take for example, Bill O'Reilly's show, *The O'Reilly Factor*," Gold writes. "[In May 2014] O'Reilly had his lowest month since 2001 in the key [demographic], with 308,000 viewers." "Yes, O'Reilly is still the No. 1 program in cable news in both total and demo viewers, averaging 2,136,000 total viewers in May. But the majority of those viewers are over the age of 55. In fact, the median age for O'Reilly is now just over 72 years old. The average Fox News viewer overall is 68.8, while the average ages of MSNBC and CNN viewers were 62.5 and 62.8, respectively." During 2015, in a scandal similar to that of NBC anchor Brian Williams, who was suspended for six months for embellishing his involvement in a news story from Iraq, O'Reilly's veracity was challenged by a list of misstatements attributed to him published on the Internet.

Problems with audience and revenues are not limited to television. Print media also appear to be dying, with only about 25 percent of the population indicating confidence in newspapers. At least 152 newspapers closed their doors in 2011 alone, and print advertising revenues fell from $49 million in 2006 to $22 million in 2012. This trend continued into 2014 with classified advertising revenues also declining.

While print advertising continued to lose revenue, media made up some of the difference with digital ad revenue, but not nearly enough. In 2012, the ratio was about fifteen print dollars lost for every digital dollar gained.

Although the decline in newspaper readership has been blamed on younger audiences deserting for electronic media, this is not the sole explanation. Total visits to newspaper websites decreased by 5 percent in 2012. The *New York Times* led all U.S. newspapers in total audience, even though it too was hemorrhaging readers, dropping from 4,442,074 in 2010 to 4,356,555 in 2012.

According to the American Society of News Editors, full-time professional editorial staffs, which peaked at 56,900 in 1989, had, by the end of 2011, fallen by 29 percent. It was estimated that by 2014 newsroom staffs would drop below 40,000.

Both print and electronic media run on the quest for larger audiences. Whichever medium has the biggest audience gets the largest revenues. But this may be a false predicate. It seems apparent from the loss of younger audiences that the sheer race for audience is not responsible for the death of the corporate mass media. A 2013 Gallup poll showed that a whopping 77 percent of those polled said they did not trust mainstream television. Only Congress came in with worse numbers, with less than 10 percent expressing any trust in the legislative branch.

Lack of trust rather than age may more fully explain the desertion of TV viewers. A 2014 Gallup poll showed Americans' confidence in the media's ability to report was at an all-time low of 40 percent. Americans belief that the corporate media present the news fully, accurately, and fairly has declined steadily from the relatively high levels of the late 1990s and the early 2000s.

According to Gallup, "Though a sizable percentage of Americans continue to have a great deal or fair amount of trust in the media, Americans' overall trust in the Fourth Estate continues to be significantly lower now than it was ten to fifteen years ago." The pollsters added that statistics showed that national elections particularly trigger skepticism about the accuracy of the news media's reports.

Unsurprisingly, Gallup reported 44 percent of Americans feel the news media are "too liberal." Only 19 percent believe they're too conservative, while 34 percent, only about one in three, say the media are "just about right" in terms of their coverage.

"The mainstream media has failed to inform us on so many levels. You can pick any day or week of the year, and observe the most trending news items littering our television screens. What you'll find is a news media that desperately holds onto any celebrity gossip for days on end, and lies through its teeth at every opportunity," wrote Joshua Krause in the *Daily Sheeple*. Speaking about the lack of coverage of the 2014 meeting of the secretive Bilderberg group in Denmark, Krause voiced the thought of many young people by noting, "They failed to report on the possibly earthshaking events that could unfold from a yearly meeting of the most powerful and influential people on earth. If they can't do that, then what are they good for?

"The truth of our world is filled with awe and wonder. They could get all the ratings they could possibly dream of, if they just told the truth. And yet, from the school textbooks of our formative years to the talking heads of our adult lives, every source of mainstream information appears to be a sanitized version of the truth, or even an outright lie," Krause added.

Many ill-informed citizens believe the untrustworthiness of the mass media stems simply from incomplete information presented by uninformed talking-head news anchors, such superficial reporting due to sloppy and credulous reporters. Those who have studied the history of media corporate ownership and control come to realize that the lack of truthful information stems from a conscious agenda of the globalist owners. This agenda includes keeping antiestablishment viewpoints away from the public and the repetitious presentation of pro-government and corporate pronouncements.

Proof of Krause's idea of truth trumping the corporate mass media came in April 2014 when Nevada rancher Cliven Bundy, along with family, friends, and supporters, stood off armed federal agents. Capping a twenty-year legal battle over grazing fees to the Bureau of Land Management (BLM) and armed with a federal court order, officers began moving Bundy's cattle off the land but were stopped by a Bundy blockade.

There was an dramatic standoff between heavily armed participants on both sides. Local law enforcement joined BLM officers,
while neighbors and supporters, including some militia members
and ex-soldiers, joined the Bundys. Local sheriff Doug Gillespie defused the situation by negotiating with Bundy and ordered the release of his cattle.

Public reaction was decidedly mixed, with some terming Bundy
a true patriot for resisting attempts by overreaching federal officials,
while others said he was promoting anarchy.

The conventional corporate mass media only covered the Bundy
story superficially and nearly always from the government's point
of view. What has been termed the "alternative media" rose to the
occasion by reporting on the story as it unfolded. The Next New
Network posted YouTube videos with updates on the situation along
with interviews with people on the scene; Pete Santilli of Guerilla-
MediaNetwork.com reported live, while CNN aired a fluff segment
on food. When it finally did report on the developing Bundy story,
CNN announced, "Federal officials say a police dog was kicked
and officers were assaulted" when live video from the scene clearly
showed canine officers siccing a dog on protesters and shoving one
woman to the ground.

Other independent journalists, such as Matt Drudge and Adam
Kokesh, reported developments as they occurred, while Alex Jones'
Infowars.com reporter Kit Daniels dug into the backstory, which
concerned Senate Majority Leader Harry Reid's alleged attempt to
put Bundy out of business in order to carry out a plan to build a
$6 billion solar facility on the property once it was in government
hands.

Unlike the accounts of the Branch Davidian deaths at Waco,
Texas, the Oklahoma City bombing, or even the account of the 9/11
attacks, during which the federal government had near-total control
over media coverage and therefore could construct false narratives
for public consumption, the Bundy story went straight to the citizenry via the alternative and social media. Such nontraditional forms
of communication are beginning to outdistance conventional journalism. Citizen journalists, armed with cell phones, are presenting

a problem for corporate mass media news. Increasingly, corporate and government officials are refusing to talk to any journalist they consider not working for a "credible" news outlet.

"The bottom line is that the mainstream media thinks you are incredibly stupid and will buy anything they say, no matter how illogical or irrational it might be," stated Mike Adams in an April 2014 article in *NaturalNews*. "What the alternative media has now proven is that the mainstream media is largely irrelevant. It matters nothing what they print or broadcast. The people who are informed know it's all lies, and the mind-numbed propaganda victims who still watch [networks] like CNN and MSNBC are irrelevant to the march of history anyway.

"Real history is being shaped, investigated and reported by the alternative media. We are the ones who have no big corporate sponsors and no million-dollar budgets, but we have the hearts and minds and passion for truth and justice that drives our work to levels of authenticity that the mainstream media can never hope to attain . . . regardless of production budgets."

"The mainstream media is on its last breath, and they are already scurrying to secure phony 'alternative news' websites in a bid to stay afloat, but the new era of news 2.0 is already here. And in this new paradigm of content consumption, reality is king," wrote Anthony Gucciardi, host of the website Storyleak.com. "The mainstream media is afraid of the new media, they are afraid of you. If one man or woman with a smartphone can change history, that is a scary thought for the political control freaks who seek to censor you at every turn. The new era of news consumption has arrived, and it's time to kick the mainstream media out for good."

Conservatives sometimes complain about the "liberal" media, but a serious look reveals that the mainstream media only tilts liberal on certain social issues such as abortion, same-sex marriage, and gun control. Otherwise, the so-called liberal mass media is only as liberal as its corporate masters allow, with stories on corporate malfeasance and corruption getting short shrift.

It has been shown that globalists spend huge sums of money manipulating media viewpoints. In early 2014, David Brock, a Democratic Party operative, revealed that his organization, Media Matters for America (MMFA), uses money from billionaire globalist George Soros to work directly with establishment journalists to influence the corporate media in an attempt to counteract the alternative and conservative media. White House visitor logs as reported in the *Daily Caller,* a twenty-four-hour news publication, showed that Brock and MMFA officials met regularly with Obama aides including Deputy Communications Director Jen Psaki and senior adviser Valerie Jarrett. "Media Matters has now been completely exposed as little more than an attack dog for the Obama administration," accused Paul Joseph Watson on the Infowars website.

Radio talk-show host and former NASA scientist Michael Rivero, a longtime media critic, summed it all up when he stated, "Most people prefer to believe that their leaders are just and fair, even in the face of evidence to the contrary, because once a citizen acknowledges that the government under which he lives is lying and corrupt, the citizen has to choose what he or she will do about it. To take action in the face of corrupt government entails risks of harm to life and loved ones. To choose to do nothing is to surrender one's self-image of standing for principles. Most people do not have the courage to face that choice. Hence, most propaganda is not designed to fool the critical thinker but only to give moral cowards an excuse not to think at all."

Elliot D. Cohen is director of the Institute of Critical Thinking: National Center for Logic-Based Therapy, and executive director of the National Philosophical Counseling Association (NPCA). Writing in *Project Censored 2014,* Cohen stated, "It would be naive to expect a government that seeks power and control over its citizens *not* [emphasis in the original] to use its influence over the corporate media in order to spread self-serving propaganda. Inasmuch as the corporate media need government to maximize their bottom line—through tax breaks, military contracts, relaxed media ownership rules, access to its officials and spokespersons, as well as other incentives and kickbacks—government has incredible power and leverage over the corporate media. Thus, instead of blaming the government for having lied to and deceived its citizens, better not to allow ourselves to be suckered into believing such propaganda in the first place. As this chapter argues, our liberties are most vulnerable to faulty thinking and best defended by sound logic."

"A contemporary dictator would not need to do anything so obviously sinister as banning the news," says Alain de Botton, author of *The News: A User's Manual.* "He or she would only have to see to it that news organizations broadcast a flow of random-sounding bulletins, in great numbers but with little explanation of context, within an agenda that kept changing, without giving any sense of the ongoing relevance of an issue that had seemed pressing only a short while before, the whole interspersed with constant updates about the colorful antics of murderers and film stars. This would be quite enough to undermine most people's capacity to grasp political reality—as well as any resolve they might otherwise have summoned to alter it. The status quo could confidently remain forever undisturbed by a flood of, rather than a ban on, news . . . when news fails to harness the curiosity and attention of a mass audience, a society becomes dangerously unable to grapple with its own dilemmas and therefore to marshal the popular will to change and improve itself."

The twenty-four-hour, seven-days-a-week news channels leave the impression that the American audience is well informed. This is not true. Airtime is filled with such a constant stream of disconnected and unprobed reports that it paints a false, even grotesque picture of the world that herds viewers into conformity.

Media critic Michael Parenti, a lecturer at a number of universities, noted that viewers are bombarded with snippets such as "fighting broke out in the region," or "many people were killed in the disturbances," or "famine is on the increase." "Many things are reported in the news but few are explained. Little is said about how the social order is organized and for what purposes. Instead we are left to see the world as do mainstream pundits, as a scatter of events and personalities propelled by happenstance, circumstance, confused intentions, bungled operations, and individual ambition—rarely by powerful class interests."

Parenti links class interest to "globalization, a pet label that the press presents as a natural and inevitable development. In fact, globalization is a deliberate contrivance of multinational interests to undermine democratic sovereignty throughout the world."

But the answer can't be just to intimidate people into consuming more "serious" news; it is to push so-called serious news outlets into learning to present important information in ways that can properly engage audiences, advises de Botton. The challenge is to have mass media outlets offer thoughtful and meaningful information—not just what happened but placed into a context including the question of why something happened and who, if anyone, benefitted.

"In the ideal news organization of the future, the ambitious tasks of contextualization and popularization would be taken so seriously that stories about welfare payments would be (almost) as exciting as those about incestuous antipodean cannibals," he opined.

With the loss of trust in the corporate mass media comes a new demand on the individual to think for him or herself and to improve thought processes to foster democracy and protect against totalitarianism.

Better understanding of the realities of the world can be achieved by not just believing the status quo, but questioning it; looking for consistency in news reports; being wary of fearmongering and media-induced stereotypes; searching for explanations and questioning all authority.

CHAPTER 23

COMING COLLAPSE?

BY THE MID-TWENTY-FIRST CENTURY, AN INCREASING NUMBER OF commentators and authors have been foreseeing American society collapsing from within, and relatively soon. They see the causes of such a collapse as numerous, varied, but also inevitable. One of the more prevalent theories involves a breakup of the financial system. Several financial forecasters in 2014 were predicting the imminent demise of the U.S. dollar and possibly the entire financial system. Some think such an eventuality might include major riots in the cities and even the imposition of martial law. As will be described later in this chapter, there is even some evidence that the federal government is preparing for just such an eventuality.

Such scary prospects are reported by commentators such as Harry Dent, who, in *The Great Depression Ahead*, predicts, "The U.S. economy is likely to suffer a minor or major crash by early 2015 and another between late 2017 and late 2019 or early 2020 at the latest."

"I think the crash of 2008 was just a speed bump on the way to the main event . . . the consequences are gonna be horrific . . . the rest of the decade [2010 to 2020] will bring us the greatest financial calamity in history," warned Mike Maloney, author and host of the *Hidden Secrets of Money* video series.

"You saw what happened in 2008–2009, which was worse than

the previous economic setback because the debt was so much higher," noted James "Jim" Rogers, chairman of Rogers Holdings and Beeland Interests, Inc. "Well, now the debt is staggeringly much higher, and so the next economic problem, whenever it happens and whatever causes it, is going to be worse than in the past, because we have these unbelievable levels of debt, and unbelievable levels of money printing all over the world."

Jeff Berwick, financial editor of the *Dollar Vigilante,* predicted, "If they allow interest rates to rise, it will effectively make the U.S. government bankrupt and insolvent, and it would make the U.S. government collapse . . . They are preparing for a major societal collapse. It is obvious and it will happen, and it will be very scary and very dangerous."

David Stockman, former director of the Office of Management and Budget under President Ronald Reagan noted, "We have a massive bubble everywhere, from Japan, to China, Europe, to the UK. As a result of this, I think world financial markets are extremely dangerous, unstable, and subject to serious trouble and dislocation in the future."

"I can tell you as someone who absolutely aced academic mathematics in my younger years that the global economy is headed for a disastrous debt collapse," avowed Mike Adams, editor of *NaturalNews,* a popular website covering health and politics that boasts more than five million visitors monthly. "Trillions of dollars of asset valuation (in derivatives) will vanish literally overnight. Widespread economic destruction will strike humanity like a thousand hurricanes hitting major population centers all across the world, all simultaneously. The timing of this is impossible to predict, but its inevitability is not. What really alarms me about all this is knowing in advance that this event will usher in a global wave of poverty and destitution that is unprecedented in all of human history. This is going to put honest, hard-working people on the streets, living in destitution, through no fault of their own. And the mere awareness of knowing this is coming causes me tremendous pain.

"Even worse, you and I can't save them all. We can only teach people to get prepared and hope they have the wisdom to listen. We cannot make their decisions for them, and we cannot alter the laws of economic reality which dictate a global day of reckoning."

Several observers of the social scene have compared the decline of the American Empire to that of ancient Rome, noting that while the Roman authorities pacified the masses with free bread and circuses filled with fighting gladiators, the American public is provided low-cost processed food and fights on television.

Internet commentator Jack Curtis, a frequent contributor to the website *American Thinker*, writes that America is running out of money "borrowed from its pressured citizens' kids and grandkids via Federal Reserve 'Quantitative Easing' games. "Nobody can live on promises forever," noted Curtis. "Stock markets and banks will shortly exhale the funny-money hot air sustaining them, interest rates will start their climb back to normal and the government will defund first, its war machine and foreign bribe programs, then its welfare beneficiaries. As with the old Romans, American military will decline, along with cradle-to-grave social welfare. And American citizens are likely to see change . . . more change than they expected that their president had in mind. Food stamps will be cut and huge salaries will disappear from pro sports. As that financial wave crests and begins to recede, America won't be an empire anymore. It will have everything it can handle just tending to its own business."

Michael T. Snyder, publisher of *The Economic Collapse Blog*, foresees a great storm coming to America in the near future in the form of a takeover by China. "Chinese acquisition of U.S. businesses set a new all-time record [in 2013], and it is on pace to absolutely shatter that record this year. Meanwhile, China is voraciously gobbling up real estate and is establishing economic beachheads all over America. If China continues to build economic power inside the United States, it will eventually become the dominant economic force in thousands of small communities all over the nation.

"And it is important to keep in mind that there is often not much of a difference between 'the Chinese government' and 'Chinese corporations.' In 2011, 43 percent of all profits in China were produced by companies that the Chinese government had a controlling interest in."

Devvy Kidd, a federal government whistle-blower, two-time congressional candidate, and author of *Why a Bankrupt America*, has written that an "outlaw" Congress has "destroyed our most important job sectors: industrial, manufacturing and agriculture via destructive, unconstitutional 'free' trade treaties, turning America into a dying service economy. It must be reversed, but Americans are going to go through hell before that can be accomplished.

"The arrogant and ignorant in the fifty state capitols have refused to implement a constitutional sound money law," she said. "So many of us warned until the (expletive) hit in August 2008, but the masses didn't listen and they still aren't listening. The very worst is closing in on us, and when it finally hits as it did in 2008, what's going to happen? Social breakdown? Yes. Food riots. I believe we'll see that in certain parts of the country because (1) people are broke and empty bellies make for angry mobs, and (2) there are serious problems with our food and water supplies."

Brandon Smith, founder of the barter network Alternative Market Project, in a 2014 article entitled "The Final Swindle of Private American Wealth Has Begun," stated, "The financial crash of 2008, the same crash which has been ongoing for years, is NOT [emphasis in the original] an accident. It is a concerted and engineered crisis meant to position the U.S. for currency disintegration and the institution of a global basket currency controlled by an unaccountable supranational governing body like the International Monetary Fund (IMF). The American populace is being conditioned through economic fear to accept the institutionalization of global financial control and the loss of sovereignty."

Such critics are not the only ones who see a bleak future for America. A recent study, sponsored in part by NASA's Goddard Space Flight Center, predicted a collapse of Western industrial civilization in the near future because of increasing income inequality along with the unsustainable exploitation of resources. According to this study, the rise and fall of civilizations is a recurrent historical cycle in which "precipitous collapses—often lasting centuries—have been quite common." The independent research project was conducted by a team of natural and social scientists under a NASA grant and was accepted for publication in the peer-reviewed Elsevier journal *Ecological Economics*, which covers both ecology and human economics.

Looking at such factors as population, climate, water, agriculture, and energy in the decline and fall of past civilizations, researchers were able to correlate their findings with the world today. They found the two crucial factors leading to collapse to be depletion of natural resources and "the economic stratification of society into Elites and Masses (or Commoners)." In words that recall the grievances publicized by the Occupy Movement, the study found fault not only with the pillaging of resources by wealthy capitalists, but also observed that "accumulated surplus is not evenly distributed throughout society, but rather has been controlled by an elite. The mass of the population, while producing the wealth, is only allocated a small portion of it by elites, usually at or just above subsistence levels."

Unsurprisingly, the study found that "commoners" are more likely to both see the abuse of resources and seek action to equalize income distribution than the wealthy elite, who are either oblivious to this "catastrophic trajectory" or "in support of doing nothing."

Dr. Nafeez Ahmed, executive director of the Institute for Policy Research and Development and author of *A User's Guide to the Crisis of Civilization: And How to Save It*, concluded, "The NASA-funded Human And Nature Dynamical (HANDY) model [namely, that wealth distribution today is unequally divided between 'haves' and "have-nots'] offers a highly credible wake-up call to governments, corporations and business—and consumers—to recognize that 'business as usual' cannot be sustained, and that policy and structural changes are required immediately."

A collapse due to social pressures may prove slow in coming. Other possibilities could be quick and varied. These include the setting off of an EMP (electromagnetic pulse) weapon, terrorist attacks in cities, a nuclear war, a national truckers strike, civil war, a cyber attack on computer systems, geophysical disasters such as the eruption of the Yellowstone caldera, and even an asteroid strike from space.

Drought conditions and lack of water also have the potential to be catastrophic, accelerating an agricultural collapse resulting in mass starvation. Even NBC News reported on a looming crisis over water being drained from the Ogallala Aquifer, a ten-million-year-old underground water source stretching from South Dakota to Texas that supplies irrigation to an eight-state agricultural region.

"The scope of this mounting crisis is difficult to overstate," wrote Brian Brown. "The High Plains of Texas are swiftly running out of groundwater supplied by one of the world's largest aquifers—the Ogallala. A study by Texas Tech University has predicted that if groundwater production goes unabated, vast portions of several counties in the southern High Plains will soon have little water left in the aquifer to be of any practical value." The worsening water situation in California as detailed previously only adds to the problem.

With all of these disturbing indicators, Alt-Market.com blogger Brandon Smith believes "a second American Revolution is inevitable."

"What frightens the establishment most, I think, is that the American people have become active participants in their own national environment once again [emphasis in the original]," Smith wrote. "At [Nevada rancher Cliven] Bundy['s] ranch, they stopped asking for mercy, they stopped begging the system to police itself, they stopped waiting for the rigged elections, and they stopped relying on useless legal avenues to effect change. Rather, they took matters into their own hands and changed the situation on the ground on their own. For oligarchy, this development is unacceptable, because one success could lead to many . . . for at least the past four years our government has been quietly maneuvering toward martial law. It's been happening for much longer if you count George W. Bush's Presidential Decision Directive 51, which has yet to be fully declassified." This directive, part of the Continuity of Operations plan, details how the executive branch agencies of the government could take control and maintain federal authority during a declared national emergency. Critics say the directive gives the president dictatorial powers and eliminates the last roadblocks to declaring martial law.

As in the past, when faced with an increasingly noncompliant citizenry, the global elite turns to war as a means of distracting the public, solidifying political control through patriotism, and damping down social movements while at the same time increasing the profits of their corporate holdings.

While even the globalists are hesitant to provoke another world war, as the massive release of nuclear weapons could spell doom for the entire planet, they are not above stimulating localized warfare across the world, particularly in the volatile Middle East.

California Internet commentator Richard Scheck noted that facts found in the corporate mass media reveal that "a Leviathan has emerged at the dawn of the new millennium reflecting the vision of *1984* and the warning of President Eisenhower to beware of the military-industrial complex." He echoed the thoughts of many by contending that factions within Western intelligence agencies control so-called terrorist groups and use them to perpetuate a "strategy of tension" and clash of civilizations to support partisan political positions.

"Psy-op [psychological operations] programs such as Operation Gladio [code name for a NATO plan to leave behind anticommunist assets in the event of a Soviet invasion of Europe during the Cold War] and reliance on paramilitary groups to continue Cold War efforts designed to defeat world communism which is currently replaced in Orwellian fashion by radical Islam (soon to be followed by China). The public is confused and manipulated by 'wag the dog' type events such as OKC [the Oklahoma City bombing], 9/11, Madrid and London. Factions within various intelligence agencies acting on behalf of vast banking, corporate and criminal (drug) enterprises exploit the rhetoric of radical groups in false-flag operations designed to terrorize the populace while conveniently shifting blame to the demonized group. This allows cabal members to acquire more power and expand their domination over all agencies of government by centralizing power (Homeland Security and the new National Intelligence Agency) . . . In this fight for the future, we are close to a tipping point where a perfect storm of tribal, economic, political, religious and environmental factors will force everyone to awaken to the crisis at hand.

"We are all increasingly becoming participants in 'war of the worlds' type scenarios as people are impacted by the consequences of peak oil, global warming, economic globalization and international terrorism. How we learn to live with the Leviathan and respond to the difficult tests that lie ahead will determine the fate of our children and the shape of the world they inherit."

Scheck recalled the Founding Fathers warning against foreign entanglements. "Those forgetting that are the real traitors to our heritage and have ruined this country in their drive for Pax Americana and global hegemony," he wrote.

Up to today, vast numbers of Americans have largely accepted the military adventures indulged by Washington because they retained a patriotic trust in their government. Such trust has succeeded so far in preventing a major societal collapse. Ever since the issue of federal dominance over the states was decided in 1865, Americans have been taught to trust their government. Yet this trust has begun to erode.

Professor Henry Giroux is an award-winning professor who taught at universities in Boston and Miami and the author of *Neoliberalism's War on Higher Education* and *Zombie Politics in the Age of Casino Capitalism*, among more than fifty titles. Giroux questions why so many citizens trust the government to protect them in the first place. He wonders, "Why should anyone trust a government that has condoned torture, spied on at least thirty-five world leaders, supports indefinite detention, places bugs in thousands of computers all over the world, kills innocent people with drone attacks, promotes the post office to log mail for law enforcement agencies and arbitrarily authorizes targeted assassinations? Or, for that matter, a president that instituted the Insider Threat Program, which was designed to get government employees to spy on each other and 'turn themselves and others in for failing to report breaches,' which includes 'any unauthorized disclosure of anything, not just classified materials.'" Some say this program was designed to turn government employees, such as your postman, into an army of snitches.

The Rutherford Institute's John W. Whitehead, in recalling how the people of Stalin's Soviet Union and Hitler's Germany blindly followed government officials, explained such blind trust in government thusly: "Unfortunately, 'we the people' have become so trusting, so gullible, so easily distracted, so out-of-touch, so compliant and so indoctrinated [to] the idea that our government will always do the right thing by us that we have ignored the warning signs all around us, [and not] asking the right questions, demanding satisfactory answers, and holding our government officials accountable to respecting our rights and abiding by the rule of law has pushed us to the brink of a nearly intolerable state of affairs . . . at least to those who remember what it was like to live in a place where freedom, due process and representative government actually meant something."

Others ask how one can trust a government that is preparing for internal strife, even a collapse of society. Because it would appear that the United States government is doing just that.

Beginning in 2008, at the outset of the financial crisis, the Department of Defense showed concern over the possibility of national collapse by funding universities to initiate studies "to improve DoD's basic understanding of the social, cultural, behavioral, and political forces that shape regions of the world of strategic importance to the U.S." This program, entitled the "Minerva Research Initiative," basically is designed to predict and prepare for social collapse across the globe to include the United States. The *Guardian* said the program was designed "to model the dynamics, risks and tipping points for large-scale civil unrest across the world, under the supervision of various U.S. military agencies."

The project will determine "the critical mass (tipping point)" of what are called "social contagions" by studying their "digital traces" in the cases of social unrest such as "the 2011 Egyptian revolution, the 2011 Russian Duma elections, the 2012 Nigerian fuel subsidy crisis and the 2013 Gazi park protests in Turkey."

The titles of projects funded by the initiative avoid words like "collapse," "rioting," and "civil war," preferring to mask this research with such headings as "Tracking Critical-Mass Outbreaks in Social Contagions," "Deterrence with Proxies," "Using New Approaches to Measure and Model State Fragility," "A Computational Assessment of Social Disequilibrium and Security Threats," and "Understanding the Origin, Characteristics, and Implications of Mass Political Movements." But the intent of the program is clear—to identify antigovernment sentiment, pinpoint any activist leaders, and devise ways to suppress government dissent. Many of the projects are geared toward foreign nations, particularly in Asia and the third world.

In 2014, Congress authorized a total budget of $17.8 million for the Minerva Initiative. However, the final program is expected to cost the taxpayers $75 million over a five-year period.

Critics of the program include the American Anthropological

Association (AAA), which complained that the Pentagon lacks "the kind of infrastructure for evaluating anthropological [and social science] research" in a manner involving "rigorous, balanced and objective peer review." In a letter to the U.S. government, the AAA stated, "Pentagon officials will have decision-making power in deciding who sits on the panels" and that "there remain concerns within the discipline that research will only be funded when it supports the Pentagon's agenda."

David Price, a professor of cultural anthropology at St Martin's University in Washington, D.C., is author of *Weaponizing Anthropology: Social Science in Service of the Militarized State.* He has been critical of the Pentagon's Human Terrain Systems (HTS) program, which is designed to embed social scientists in military field operations, applying society-altering theories to military "nation building" activities. Price notes the HTS training scenarios adapt counterinsurgency tactics used in Iraq and Afghanistan for use in the USA, "where the local population is seen from the military perspective as threatening the established balance of power and influence, and challenging law and order."

Such public studies concern citizens already aroused over the militarization of police and the federal government's stockpiling of arms and ammunition. One particularly disturbing study in 2013, entitled "Who Does Not Become a Terrorist," equated peaceful activists with armed militants. Study material stated, "This project is not about terrorists, but about *supporters* of political violence."

Guardian reporter Nafeez Ahmed queried Pentagon officials: "Activism, protest, 'political movements' and of course NGOs are a vital element of a healthy civil society and democracy—why is it that the DoD is funding research to investigate such issues?" Ahmed said he received no clear answer.

Rutherford Institute founder Whitehead questioned the militarization of government agencies not known for firefights, viewing it as evidence that the federal government is preparing for a societal collapse. He noted a buildup in recent years of SWAT teams within non-security-related federal agencies such as Department of Agriculture, the Railroad Retirement Board, the Tennessee Valley Authority, the Office of Personnel Management, the Consumer Product Safety Commission, the U.S. Fish and Wildlife Service, and the Education Department. He further asked why at least seventy-three federal agencies under the command of Homeland Security or the Justice Department require approximately 120,000 full-time armed officers with arrest authority.

"What's with all of the government agencies stockpiling hollow-point bullets? For example, why does the Department of Agriculture need .40-caliber semiautomatic submachine guns and 320,000 rounds of hollow-point bullets? For that matter, why do its agents need ballistic vests and body armor?" Whitehead asked. "Why does the Postal Service need 'assorted small arms ammunition'? Why did the DHS purchase 1.6 billion rounds of hollow-point ammunition, along with 7,000 fully automatic 5.56x45mm NATO 'personal defense weapons' plus a huge stash of 30-round high-capacity magazines? That's in addition to the FBI's request for 100 million hollow-point rounds. The Department of Education, IRS, the Social Security Administration, and the National Oceanic and Atmospheric Administration, which oversees the National Weather Service, are also among the federal agencies which have taken to purchasing ammunition and weaponry in bulk."

In mid-2015 a multi-state military exercise called Jade Helm 15 involved not only special operations units such as the Army's Green Berets, Navy SEALs, and Air Force Special Ops but also law enforcement agencies. Participants were to practice infiltrating both urban and rural areas to identify and detain citizens thought to be resistant to government demands. The fact that Texas, Utah, and lower California are listed as "hostile states" prompted concerns that the exercise not only violated the 1878 Posse Comitatus Act restricting the use of the military to police U.S. citizens but was a forerunner to martial law.

In early June 2014, an incident took place in Houston that might presage a coming collapse in America.

When a twenty-three-year-old woman was killed in a traffic accident, bystanders looted her car of groceries even as her two injured children sat inside it. Police said the woman lost control of her Toyota

4Runner after being clipped by another car while backing out of a private drive. She hit a tree and was pronounced dead at the scene. Her sons, ages four and six, were in the backseat and were transferred to a hospital with broken bones but no life-threatening injuries.

Witnesses told police they saw people steal groceries out of the dead woman's SUV. Although it was not clear if the looting took place before or after the woman's body and the children were removed from the wreck, it is most probable that the theft occurred before emergency personnel arrived.

One witness told newsmen, "Why would you take somebody's stuff who got hit by a car? That's crazy, that's mean." But a nearby resident, Savannah Roberts, said she was not surprised by the looting. "There is a lot of people you just can't trust," she said. "I've seen worse in this area."

What worries some even more than the scarcity of food is the thought that some multinational corporation—Monsanto comes to mind—might one day have the power to shut down the world's food supply without notice.

Already having been the object of worldwide protests over toxic chemicals such as Agent Orange, PCBs (polychlorinated biphenyls), and dioxin, many critics are fearful of Monsanto's reach for control of the world's food supply through its proprietary seeds and GMOs. Today, more than 40 percent of all U.S. crop acreage use Monsanto products and the company owns more than 1,600 patents on seed, plant, and other related commodities.

Food & Water Watch, a nonprofit group with fifteen U.S. offices that is dedicated to safe, accessible, and sustainable food and water, produced a paper in 2013 entitled "Monsanto: A Corporate Profile." According to this paper, in the United States alone, nearly all (93 percent) of soybeans and four-fifths (80 percent) of corn were grown with seeds containing Monsanto GMOs. Monsanto's leading products include Roundup and Harness herbicides, DeKalb corn seeds, Asgrow soybean seeds, Deltapine cotton seeds, Seminis and De Ruiter vegetable seeds, and insect repellent Smartstax corn and Bollgard cotton.

"The company's power and influence affects not only the U.S. agricultural industry, but also political campaigns, regulatory processes and the structure of agriculture systems all over the world," noted the paper.

Supreme Court rulings in 1980 and 2001 have allowed the patenting of living organisms, leading to the development of what are termed "Terminator seeds," crop seeds that will not reproduce the next growing season. What was once a freely exchanged and renewable food source has been privatized and monopolized. "In less than three decades, a handful of multinational corporations have engineered a fast and furious corporate enclosure of the first link in the food chain," stated the Food & Water Watch report.

Although Monsanto insists that it would never commercialize Terminator seeds, depriving farmers of new crop seeds or simply producing seeds not engineered to produce a full crop could cause worldwide famine. Such a dastardly scheme also would fit quite nicely with the broader globalist population control agenda. Any crisis in the food supply could certainly initiate chaos of such magnitude that the public would cry out for martial law, especially in the large cities where hungry mobs of looters could overrun police forces. Most citizens today cannot imagine such a possibility because they are psychologically affected by a phenomenon known as normalcy bias, whereby people fail to recognize or underestimate the possibility of disaster. Most people tend to believe that whatever they experience on a day-to-day basis is "normal" and that things will stay that way. Such bias prevents them from considering the ramifications of current trends.

"As a practical example, most of us suffer under the normalcy bias delusion that when you turn on the faucet in the kitchen, water will always come out," explains Mike Adams. "We've seen this happen so many times that we now take it for granted and believe it will always happen, almost as if by magic. Even though humans living two hundred years ago would have been shocked to see clean water coming out of a kitchen faucet, today we are shocked if it doesn't come out . . . ! That's normalcy bias."

Obviously, if the citizens of the death culture hope to survive and flourish in the future, both their mind-set and behavior must change.

CHANGING THE GAME

WHAT CAN CITIZENS DO TO PREVENT THE POSSIBILITY OF A FUTURE of privation and tyranny? What can be done on an individual level to make life better, to change the game? After viewing the egregious missteps of modern American society, what might be done to advance ideas and philosophies that could correct past mistakes? What can engender a more peaceful and prosperous nation?

As author Alain de Botton notes, the problem with facts is not that we need more of them, but that we don't know what to do with the ones we have. The news media spews out an avalanche of facts each day, but what do these facts really mean? "What should be laudable in a news organization is not a simple capacity to collect facts, but a skill—honed by intelligent bias—at teasing out their relevance," writes de Botton. "We need news organizations to help our curiosity by signaling how their stories fit into the larger themes on which a sincere capacity for interest depends." In today's topsy-turvy America, when considering what needs to be fixed and how, one faces a number of puzzling contradictions.

A list of conundrums passed around the Internet in 2014 was repeated by commentator and a decorated former army lieutenant colonel Allen West in his website column. Here are six conundrums of socialism that West says "pretty much [sum] up the USA in the twenty-first century":

1. America is said to be capitalist and greedy—yet half of the population is subsidized.
2. Half of the population is subsidized—yet they think they are victims.
3. They think they are victims—yet their representatives run the government.
4. Their representatives run the government—yet the poor keep getting poorer.
5. The poor keep getting poorer—yet they have things that people in other countries only dream about.

6.   They have things that people in other countries only dream about—yet they want America to be more like those other countries.

West went on to note other contradictions in American society. He pointed out that Americans are advised to NOT judge ALL Muslims by the actions of a few lunatics, but we are encouraged to judge ALL gun owners by the actions of a few lunatics [emphasis in the original].

He said it seems we constantly hear about how Social Security is going to run out of money. "How come we never hear about welfare or food stamps running out of money?" asked West. "What's interesting is the first group worked for their money, but the second didn't."

Finally, West, a veteran of the conflict in Iraq, pondered, "Why are we cutting benefits for our veterans, no pay raises for our military and cutting our army to a level lower than before WWII, but we are not stopping the payments or benefits to illegal aliens? Am I the only one missing something?"

Maybe, just maybe, the answer rests with smaller government. Change must begin at the local level as the corruption and ambitions in Washington are beyond quick remedy. Initially, there must be changes in both public and private attitudes toward government, the corporations, and toward each other. New ideas for the better use of energy, communications, transportation, and health care must be developed.

People in America must understand that while no one can prevent banks from collapsing or ensure that a home can be sold for what it originally cost, there are some measures that can be taken to provide some protection in a chaotic future. A good rule of thumb might be "hope for the best but plan for the worst."

If at all possible, move out of the city. Buy farmland with access to water. A farm with a year-round spring may be worth more than gold in the coming years.

Those hoping for security in the U.S. dollar or with savings accounts, stocks, and bonds could lose their shirts in the event of a financial system meltdown. Diversification may be the key to survival. This may be the time to trade in paper dollars for real assets such as land, tractors, medical supplies, precious metals, ammo, food, and so on.

One California family has provided an example of what can be accomplished by changing priorities. At a time when the federal government is run by multinational corporations and the general public is zombified by processed food and TV, this family has worked to become largely self-sufficient by transforming a small backyard garden into a productive microfarm.

Jules Dervaes, along with his son Justin, and his two daughters, Anais and Jordanne, live in a 1,500-square-foot-bungalow on one-fifth of an acre on the edge of Los Angeles. In their small garden, they grow 350 different vegetables, herbs, fruits, and berries. The sustainable plot is complete with chickens, ducks, rabbits, goats, and honey bees. For two years in a row they were able to produce six thousand pounds of food. Hailed as one of the most independent family units in the country, the Dervaeses have progressively reduced their environmental impact and provided a sterling model for living sustainably and simply in an urban setting.

Ninety percent of their vegetarian diet comes from the homestead and two-thirds of their energy is solar. Biodiesel fuel is made from used vegetable oil. Their radio operates by a hand crank. While they consume most of the food they produce, the Dervaeses sell any excess to local individuals and businesses. Profits are used to purchase basics like flour and rice. Dervaes says his family has demonstrated how individuals need not rely on a centralized authoritarian system to live a productive and rewarding life. "Government can't do it and corporations won't do it," he explains, adding with some humor that his family is "in danger of being free."

More and more people are joining the Dervaes family and breaking with the status quo of America. They are moving out of congested cities and buying small farms, returning to the land to raise their families and their own healthy food. Those who cannot leave the city are becoming self-sufficient by cultivating backyard gardens.

Still others are shopping at farmers markets and turning to organic products in an effort to avoid the deficiencies of processed, mass-produced food. Society cannot long survive when food is no longer nutritious or when it contains poisonous substances. Supporting local markets not only helps struggling farmers but guarantees a healthy diet. In many areas food cooperatives are providing nutritious foods to members. If no food co-op is near you, form one by making contact with producers in your area.

As the Boy Scout motto states, "Be prepared." This does not mean that everyone should become a survivalist with massive stockpiles of food, water, and ammunition. It only means that any prudent person should hope for the best but prepare for the worst.

Even the federal government, in the form of the Federal Emergency Management Agency (FEMA), recommends that every family have a Basic Emergency Kit, also known as a bug-out bag. The FEMA website encourages everyone to be prepared for an emergency situation. "You may need to survive on your own after an emergency," acknowledges the site. "This means having your own food, water and other supplies in sufficient quantity to last for at least 72 hours. Local officials and relief workers will be on the scene after a disaster but they cannot reach everyone immediately. You could get help in hours or it might take days. Additionally, basic services such as electricity, gas, water, sewage treatment and telephones may be cut off for days or even a week, or longer. Your supplies kit should contain items to help you manage during these outages."

FEMA recommends accumulating the following basic supplies:

- Water, one gallon of water per person per day, for drinking and sanitation
- Food, at least a three-day supply of nonperishable food
- Battery-powered radio and a NOAA weather radio with tone alert, and extra batteries for both
- Flashlight and extra batteries
- First-aid kit
- Whistle to signal for help
- Infant formula and diapers, if you have an infant

- Moist towelettes, garbage bags and plastic ties for personal sanitation
- Dust mask or cotton T-shirt, to help filter the air
- Plastic sheeting and duct tape to shelter-in-place
- Wrench or pliers to turn off utilities
- Can opener for food (if kit contains canned food)

FEMA also advises citizens in cold-weather climates to prepare for low temperatures with sleeping bags, coats, and other warm clothing.

Over and above basic survival preparation, thoughtful people should plan for a situation in which medical services are either over-crowded or simply unavailable.

Instruction books on first aid and home and herbal remedies should be kept on hand. If a family member requires prescription medicine, a supply should be kept in reserve. Citizens must be prepared to take responsibility for themselves and their family rather than count on help from the medical and pharmaceutical establishments.

For those desiring to take control over their own health, the website naturalcuresnotmedicine.com points out that today there exists an "almost limitless library, open twenty-four hours a day, seven days a week, to anyone with a passion for reading! Never has so much free material been available." Today, there are more than a hundred websites filled with a wide variety of e-books, ranging from literary classics to how-to instructions.

Robert David Steele, a former deputy director of Marine intelligence and CIA case officer, advocates open-source intelligence, that is, information derived from public sources. Steele founded the Open Source Solutions Network Inc. and later the nonprofit Earth Intelligence Network, which supports the *Public Intelligence Blog*. Americans today, according to Steele, are lacking in public intelligence, in knowing what one needs to know in order to make honest decisions for the good of all rather than corrupt decisions for the good of the few due to distraction and misdirection by the corporate mass media.

Steele, in the spirit of the "Be Prepared" says, "I'm a former spy and I believe we still need spies and secrecy, but we need to redirect the vast majority of the funds now spent on secrecy toward savings and narrowly focused endeavors at home . . . Believe it or not, 95 percent of what we need for ethical evidence-based decision support cannot be obtained through the secret methods of standard intelligence practices. But it can be obtained quite openly and cheaply from academics, civil society, commerce, governments, law enforcement organizations, the media, all militaries, and non-governmental organizations. An Open Source Agency, as I've proposed it, would not just meet 95 percent of our intelligence requirements, it would do the same at all levels of government and carry over by enriching education, commerce, and research—it would create what I called in 1995 a 'Smart Nation.'"

According to Steele, most of our problems today can be traced to unilateral militarism, virtual colonialism, and predatory capitalism, all based on force and lies and encroachment on the common good. "The national security state works for the City of London and Wall Street—both are about to be toppled by a combination of Eastern alternative banking and alternative international development capabilities, and individuals who recognize that they have the power to pull their money out of the banks and not buy the consumer goods that subsidize corruption and the concentration of wealth. The opportunity to take back the commons for the benefit of humanity as a whole is open—here and now."

Just such a move may have begun in 2015 when fifty-seven nations joined the China-led Asian Infrastructure Investment Bank (AIIB) over the objections of the U.S. government. The AIIB was expected to compete with the Anglo American–run World Bank and International Monetary Fund.

Noting that during the past few decades, populations who had spent centuries evolving away from slavery were reduced to marketable commodities by the industrial era. In his 2012 book, *The Open Source Everything Manifesto*, Steele said communities must reject "concentrated illicitly aggregated and largely phantom wealth in favor of community wealth defined by community knowledge, community sharing of information, and community definition of truth derived in transparency and authenticity, the latter being the ultimate arbiter of shared wealth."

Steele joined many others today in calling for an end to national diktat and instead the emergence of bottom-up clarity, diversity, integrity, and sustainability. Across America, a growing number of citizens feels the federal government is out of control. True freedom and liberty must begin with individuals at the local level. "Individual towns across the USA are now nullifying federal and state regulations—for example gag laws on animal cruelty, blanket permissions for fracking," noted Steele. "Top-down power has failed in a most spectacular manner, and bottom-up consensus power is emergent. 'Not in my neighborhood' is beginning to trump 'Because I say so.' The one unlimited resource we have on the planet is the human brain—the current strategy of one percent capitalism is failing because it is killing the Golden Goose at multiple levels. Unfortunately, the gap between those with money and power and those who actually know what they are talking about has grown catastrophic. The rich are surrounded by sycophants and pretenders whose continued employment demands that they not question the premises."

Another thing individuals can do is communicate with local news media. Let them know when they do something good for the community and, more importantly, let them know when you are displeased with the coverage, or lack thereof, of a news event. Send them important news items from the alternative media and ask why they are not covering such important topics. Hold them to their word when they boast they have the best news coverage in town.

Write letters to your political representatives, who are always concerned about getting your vote. One or two letters may not sway a politician beholden to corporate contributions, but an avalanche of letters and calls can often make a difference in their voting record.

In politics, it is a sad but unalterable fact that no one can get elected to office without telling lies. In America today it appears no one can be elected at the national level by telling the truth, as witnessed by the failed campaigns of Ron Paul and Ross Perot. So the intelligent voter must disregard what any office seeker may say.

Merely watch what they do, how they vote. If they vote to your satisfaction, then vote them back into office. If not, vote them out and try someone new, regardless of party politics. This is not a revolutionary idea, it is the way the system is supposed to work.

America must move away from the political party system. Political ideologies have exerted a devolutionary influence on us. As Internet blogger Paul Rosenberg explains, "They make us harsher, angrier, easier to manipulate, and almost impossible to reason with. In almost every way that can be examined, they're bad for us," he wrote. "The great problem is that people think they're required to have political stances. This is a tremendously damaging and false belief, but the Western world is currently addicted to it. In our time, Politics is Almighty. The truth, however, is that we are more than capable of examining the world and coming to conclusions without the mental partnership of a political ideology." In other words, a free and thinking person should not have to declare him or herself a Democrat or a Republican when expressing political opinions. In fact, more and more people are realizing that the two parties are simply two wings of America's one and only political entity—the War Party.

Both Democrats and Republicans, which have now split into the more correct labels of liberals and conservatives, allow heated debates over topics such as same-sex marriage, abortion, and gun control. But when it comes to foreign policy and particularly issues involving the corporations, the banks, or the military, there is no difference between the two parties.

Rosenberg noted that once the political process is completed and laws are passed, people have a binary choice: either obey or be punished. "At the end of every political process are armed men, violently enforcing it. That's barbaric, and it's ugly," he noted. "The truth is that we really don't need those embittering ideologies. And if we ever really did need them, we've outgrown them. As best I can see it, the truth is that political ideologies make people consistently barbaric and ugly. They make them worse. Our lives would be improved if we dropped them." For a clear illustration of how a public nonviolent revolution might succeed, one need only look at the tiny island nation of Iceland. Since 2008, the corporate mass media in America has provided lavish coverage of the American financial meltdown, the "Arab Springs" in Egypt and Syria, the "Occupy" movement, and the Sunday-morning talking heads go on and on about secretive organizations such as the Trilateral Commission and the Council on Foreign Affairs preaching financial austerity. But rarely mentioned are events in Iceland, which may provide a blueprint to regaining popular control over a government and financial system.

When the financial meltdown of 2008 began, the primary banks in Iceland were nationalized and it was decided not to pay the debts created by Great Britain and Holland's questionable policies. In two separate referendums, Icelandic citizens voted that Iceland should not have to repay foreign creditors the money they lost when Icelandic banks defaulted. Elections were held and the entire government was replaced. The new regime proposed to repay the debts over a period of fifteen years at the low interest rate of 5.5 percent. In 2010, after public demonstrations, the government initiated an investigation and many high-level bankers and executives were arrested for their part in the financial crisis. Many soon left the country. A new constitution based on the Danish Constitution was written and a constitutional assembly called, composed of candidates whose only qualifications were that they were adults and had the support of thirty persons.

Many notable economists hailed the move. Columbia University economics professor and Nobel Prize winner Joseph Stiglitz wrote, "Iceland did the right thing by making sure its payment systems continued to function while creditors, not the taxpayers, shouldered the losses of banks."

Paul Krugman of the *New York Times* agreed, writing, "In a nutshell, Ireland has been orthodox and responsible—guaranteeing all debts, engaging in savage austerity to try to pay for the cost of those guarantees, and, of course, staying on the euro. Iceland has been heterodox: capital controls, large devaluation, and a lot of debt restructuring—notice that wonderful line from the IMF, above, about how 'private sector bankruptcies have led to a marked decline in external debt.' Bankrupting yourself to recovery! Seriously. And guess what: heterodoxy is working a whole lot better than orthodoxy."

Today, despite some setbacks, Iceland's economy appears strong and unemployment is much lower than it is in Ireland, Greece, or Portugal. Iceland's ordeal exemplifies the conundrum facing America—you can have the privatization of banks and other businesses and risk corruption and monopolies, or you can have centralized government control that also might not work for the benefit of the public. Yet questions remain as to why the Iceland experience was not widely or clearly presented in the corporate mass media. A contributor to CNN's *iReport* asked, "Have we been informed of this through the media? Has any political program in radio or TV commented on this? No! The Icelandic people have been able to show that there is a way to beat the system and have given a democracy lesson to the world."

One basic problem in America is that more people are voting for a living than working for a living. One 2010 study showed that 60 percent of Americans were receiving more in government benefits than they paid in taxes. This number, of course, includes government pensions, Social Security, and Medicare payments that come from contributions made during a person's lifetime and are correctly termed entitlements. These people earned this money. Yet according to the U.S. Department of Commerce, almost 47 million Americans received food stamps, though actually most now receive government-issued debit cards called Electronic Benefit Transfers (EBTs), and another 5.6 million receive unemployment insurance.

According to U.S. Census Bureau statistics released in mid-2014, almost 110 million Americans—more than one-third of the total population—were living on some form of government assistance. This number included more than 51 million receiving food stamps and 83 million collecting Medicaid, with some collecting from more than one program, and does not reflect the more than 5 million persons receiving government retirement benefits.

Consider that this same census bureau reported that the number of full-time private-sector workers, in other words, those American who got up and went to a job every morning, numbered only 86.4 million. Of a total of 103 million full-time, year-round workers, almost 17 million work for the government. This includes 12.5 million who worked for state and local governments and 4 million federal employees, all of whom are paid by tax money from the public treasury.

Is doesn't take a math whiz to realize that a nation with only 86 million full-time private-sector workers cannot sustain 110 million living off government benefits. A collapse appears imminent.

Such statistics support the words of the French historian Alexis de Tocqueville, who, after traveling in the U.S. in the early 1800s, wrote, "The American Republic will endure until the day Congress discovers that it can bribe the public with the public's money."

Anyone who has ever done his own taxes realizes that he cannot pay out more money than he has received for very long. This is why a growing number of citizens believe that people receiving government subsistence—other than entitlements earned by years of work—should not be able to vote. They see this as a clear conflict of interest, as a welfare recipient's vote will always go to whoever promises them the most from public funds. Some welfare critics believe an application for government benefits should be a voluntary renunciation of the right to vote.

Alfred W. Evans, in a November 18, 2010, letter to the editor of the *Waco* [Texas] *Tribune-Herald* reflected the views of many who wish to change the current American system when he wrote: "Put me in charge of food stamps. I'd get rid of Lone Star cards; no cash for Ding Dongs or Ho Ho's, just money for fifty-pound bags of rice and beans, blocks of cheese and all the powdered milk you can haul away. If you want steak and frozen pizza, then get a job.

"Put me in charge of Medicaid. The first thing I'd do is to get women Norplant birth control implants or tubal ligations. Then we'll test recipients for drugs, alcohol, and nicotine and document all tattoos and piercings. If you want to reproduce [and] use drugs, alcohol, smoke, or get tats and piercings, then get a job.

"Put me in charge of government housing. Ever live in a military barracks? You will maintain your property in a clean and good state of repair. Your 'home' will be subject to inspections anytime and possessions will be inventoried. If you want a plasma TV or an Xbox 360, then get a job and your own place.

"In addition, you will either present a check stub from a job each week or you will report to a 'government' job. It may be cleaning the roadways of trash, painting and repairing public housing, whatever we find for you. We will sell your twenty-two-inch rims and low-profile tires and your blasting stereo and speakers and put that money toward the 'common good.'

"Before you write that I've violated someone's rights, realize that all of the above is voluntary. If you want our money, accept our rules. Before you say that this would be 'demeaning' and ruin their 'self-esteem,' consider that it wasn't that long ago that taking someone else's money for doing absolutely nothing was demeaning and lowered self-esteem.

"If we are expected to pay for other people's mistakes we should at least attempt to make them learn from their bad choices. The current system rewards them for continuing to make bad choices."

The biggest roadblock to making positive changes in America comes from the fact that such changes will require a change of laws, which, in turn, means congressional action.

Yet this comes at a time when polls show the lowest public opinion of that body as a consequence of the polarization of the political parties. No national lawmaker seems capable of solving public issues through rational, thoughtful debate and compromise.

All national politicians can talk about today is democracy . . . we have to defend democracy, save democracy, and bring democracy to people in foreign lands. Citizens fail to understand that the definition of democracy is simply rule by the majority, and the clearest example of democracy in action is a lynch mob. The Founding Fathers did not leave us a pure democracy; they created a democratic republic ruled by laws and checks and balances.

Only after a fair trial—complete with legal representation, a jury of peers, and the opportunity to challenge the witnesses and evidence offered by the prosecution—can a defendant be found guilty and lynched.

The wealthy elite today constantly extol the benefits of democracy and majority rule, because they easily manage the masses by their ownership of a mere handful of transnational media corporations. A few critics are allowed to host alternative talk shows and websites to give the impression of First Amendment freedom. The globalists don't mind, for they still control the masses through ownership of the corporate media, and, after all, majority rules. And few realize that the media control and the polarization taking place today has been the conscious agenda of the globalists, who consider the United States as a not-so-profitable branch of the world economy they seek to control.

It is time for each thoughtful and concerned citizen to apply his or her own talents to solving the problems of America. It is too late to correct the abuses and excesses of the federal government, especially considering the problematic use of computer voting machines, which have proven to be so susceptible to hacking and manipulation and the increasing control of Congress by corporations and special interests.

So any meaningful change must begin at the personal level and work upward through local governments. With the help of neighbors and like-minded citizens, local city councils, county commissioners, and school boards should be packed with persons both aware of and concerned with the issues involved in serving and protecting the general public. Local leaders answerable to the community could turn the tide of an overreaching corporate-controlled federal bureaucracy and better manage local services.

The primary point is that each individual must begin to take personal responsibility for himself and his family. Robert David Steele reminds us that there is a wealth of open-source material to guide the individual truly seeking knowledge in whatever field, whether it is how to start a business or how to raise chickens. The time for trusting a corporate-run government to take care of every need has passed.

A 2014 Pennsylvania case is one example of successful prevention of overreach from government bureaucrats. In that case, Sheriff Carl Nace refused to hand over the names and addresses of concealed-weapon carriers to three county auditors who claimed they needed them to properly audit the sheriff's office. Nace said such a request was unwarranted for audit needs and conflicted with a statute making it a felony to disclose the names. The auditors sued the sheriff, but the judge sided with Sheriff Nace and dismissed the lawsuit. The Oath Keepers organization, the Perry County Commissioners, the Pennsylvania Sheriffs Association, the Prince Law Firm, and a large number of Pennsylvania residents supported Sheriff Nace.

"Sheriffs have a unique role in law enforcement. Unlike their counterpart, the police chief, they are elected by the people. That fact makes them accountable *to the people* [emphasis in the original]. It forms a special bond of trust between the sheriff and his constituents," notes Robert Hase with Elias Alias on the Oath Keepers website. "Sheriff Nace has decided to do what is right. He is honoring his oath to the people as required by the Constitution."

Andrew Martin, editor of Oneness Publishing and author of the 2011 book *One—A Survival Guide for the Future,* noted, "For the most part we live in a state of separation, we create thoughts and scenarios in our minds that do not represent reality. We create these realities that go toward deluding ourselves. In this world of constant struggle, we suffer, causing ourselves pain, frustration and isolation. To live free and enlightened is to release ourselves from these self-inflicted negative states of mind. We are free to do this at any time we choose, it is merely a matter of choice."

Unlike Sheriff Nace, most people feel impotent to stand against the totalitarianism of big government and the police state. "I'm only one person, what can I do about all this?" is a common refrain. Taking charge of one's own power and destiny may seem like a daunting challenge, but recent scientific breakthroughs support the idea that each individual has more capacity to change her reality than has been conditioned into her by education and religious authorities.

Today, it is becoming known that individuals can effect change within themselves by altering their thinking and emotions. Human DNA can be reprogrammed through words and emotions without cutting and replacing genes. Apparently, human DNA is a biological Internet and superior in many aspects to the artificial one, according to German authors Grazyna Fosar and Franz Bludorf, writing on the Wake Up World website. "Recent research directly or indirectly claims to explain phenomena such as clairvoyance, intuition, spontaneous and remote acts of healing, self-healing, affirmation techniques, unusual light (auras) around people, the mind's influence on weather patterns and much more," they wrote.

"Esoteric and spiritual teachers have known for ages that our body is programmable by language, words and thought. This has now been scientifically proven and explained," wrote Fosar and Bludorf.

Today, more and more children are being born with access to such DNA consciousness. They exhibit telepathic and clairvoyant abilities and develop a group consciousness. "Researchers think that if humans with full individuality would regain group consciousness, they would have a god-like power to create, alter and shape things on Earth! And humanity is collectively moving toward such a group consciousness of the new kind," wrote Fosar and Bludorf.

The belief that humans can alter their emotions and physical body through sheer willpower was also supported by the book *The Biology of Belief: Unleashing the Power of Consciousness, Matter, and Miracles* by former medical school professor and research scientist Dr. Bruce H. Lipton. Termed "groundbreaking" by critics, Lipton's book demonstrates how human cells receive and process information. The work of Lipton and other scientists shows that human DNA is controlled not through biology but through energetic signals from outside the cell, including a person's negative and positive thoughts. This means the physical body can be altered by changing the person's thoughts and emotions.

This is where science and spirituality begin to blend. According to Dr. Andrew Newberg, a professor of radiology, psychology, and religious studies at the University of Pennsylvania and cofounder of its Center for Spirituality and the Mind, "A large body of science shows a positive impact of religion on health. The way the brain works is so compatible with religion and spirituality that we're going to be enmeshed in both for a long time."

But such advances in thought, perception, and the ability to alter perceptions leading to alterations in one's physiology must be based on truth.

Those who follow the teaching of the Bible should notice that no fewer than three times in the New Testament, Jesus tells his disciples that all sins may be forgiven, even blasphemy against him (Matthew 12:31–31; Mark 3:28–29; Luke 12:10), but the one sin that can never be forgiven is blasphemy against the Holy Spirit.

The definition of the Holy Spirit is found in John 15:26 (Living Bible): "But I will send you the Comforter—the Holy Spirit, the source of all truth."

The solution to most of the world's problems may be found in the spirit of truth.

Truth is eternal. The whole universe is truth. It simply exists. Galaxies, suns, and planets just exist. Animals can love or hate but they cannot practice deceit. Only humanity, with our freewill, can choose to be deceitful. Only humans can speak untruths when they know better. So to speak against the Holy Spirit is to distort, deform, or deny truth.

This should be seriously considered by those religious fundamentalists who have allowed truth to be distorted and misused in America, who have blindly followed their leaders into a society geared toward war, death, and debilitation.

Since the Vietnam War, with government promises of "light at the end of the tunnel," to the Bush administration's warning that Saddam Hussein was within weeks of striking with "weapons of mass destruction," the American public has been pushed from pillar to post by government lies, half-truths, and demagoguery.

In 2014, President Barack Obama declared success in Iraq by withdrawing U.S. troops, only to send soldiers back midyear to protect American personnel as radical Sunni jihadist insurgents of the Islamic State of Iraq and Syria (al-Sham) (ISIS) overran many key areas. A June 2014 Gallup poll indicated that the majority of respondents viewed Obama's decision to send three hundred Special Forces advisers to Iraq as only a preliminary step in sending larger numbers of troops back to that strife-ridden country.

The mid-2014 poll, which gave Obama a six-year-low approval rating on foreign policy, showed that while most respondents thought the insurgents would conquer Iraq if the U.S. didn't return, a majority indicated they thought it more important to keep our troops out of Iraq than to stop the fighting. But, as witnessed by the 2003 invasion of Iraq, which was presaged by some of the largest anti-war demonstrations ever recorded, the will of the majority doesn't seem to alter decisions made by the corporate globalists that run the American Empire.

A feeling of impotence and cynicism in the face of globalist control was reflected in a mid-2014 Gallup poll, which gauged Americans' eroding confidence in their leaders. Public confidence in all three branches of the federal government had fallen to record lows. Only 29 percent expressed confidence in the presidency, down 7 percent from a previous rating of 36 percent. Discontent was also registered for the other two branches of government, with the Supreme Court gathering a 30 percent approval rating, while Congress remained in the single-digit column with only 7 percent expressing confidence in that institution.

The failures of centralized government aside, many feel it is long past time for the American people to cast off the blinders imposed on them by the corporate mass media and view the reality of their death-dealing society, to truly move toward a future that values life over death. Be warned, this could lead to a new and shocking worldview, as many may learn that the much-discussed "New World Order" is simply the Old World Order. The means of exercising power are the same; only the technologies have changed. The caesars and kings of yesterday became the Robber Barons of the nineteenth and twentieth centuries, who in turn have become the corporate owners of today. These self-styled globalists believe themselves to be more enlightened, entitled by heritage, and therefore, more worthy than others to rule the world.

And through their ownership of the multinational corporations that control governments and even our food, water, and pharmaceuticals, they are drawing immense profits even as they poison and sicken whole populations in their pursuit of depopulation.

The globalist-instigated American culture of death must be turned into a culture of life.

ACKNOWLEDGMENTS

THE AUTHOR WOULD LIKE TO GRATEFULLY ACKNOWLEDGE THE IN-numerable persons—authors, bloggers, journalists, talk show hosts, whistle-blowers, demonstrators, and activists—who have researched and addressed the insidious inroads into modern life made by the wealthy elite globalists. Thanks to their courageous work a growing segment of the population today has begun to see the struggle against a corporate New World Order as much more than political controversy—but rather self-defense. Sincere thanks in the making of this book goes to Maritha Gan, Thomas Ruffner, as well as to the astute editing of Nick Amphlett and Henry Ferris of HarperCollins Publishers, and also my forbearing family, Carol, Cat, Moxy, and Jackson.

Appendix:
Bilderberg Attendees

A published list of Bilderberg attendees at its 2014 meeting in Copenhagen, Denmark. Their country and position include:

Austria
Oscar Bronner, Publisher, *Der STANDARD* Verlagsgesellschaft m.b.H.
Rudolf Scholten, CEO, Oesterreichische Kontrollbank AG

Belgium
Etienne Davignon, Minister of State
Thomas Leysen, Chairman of the Board of Directors, KBC Group

Canada
W. Edmund Clark, Group President and CEO, TD Bank Group
Brian Ferguson, President and CEO, Cenovus Energy Inc.
Jason T. Kenney, Minister of Employment and Social Development
Heather Munroe-Blum, Professor of Medicine and Principal (President) Emerita, McGill University
Stephen S. Poloz, Governor, Bank of Canada
Heather M. Reisman, Chair and CEO, Indigo Books & Music Inc.

China
Yiping Huang, Professor of Economics, National School of Development, Peking University
He Liu, Minister, Office of the Central Leading Group on Financial and Economic Affairs

Denmark
Flemming Besenbacher, Chairman, The Carlsberg Group
Ulrik Federspiel, Executive Vice President, Haldor Topsøe A/S
Steffen Kragh, President and CEO, Egmont
Søren-Peter Olesen, Professor; Member of the Board of Directors, The Carlsberg Foundation
Jørgen Huno Rasmussen, Chairman of the Board of Trustees, The Lundbeck Foundation
Henrik Topsøe, Chairman, Haldor Topsøe A/S

Finland

Matti Alahuhta, Member of the Board, KONE; Chairman, Aalto University Foundation
Matti Apunen, Director, Finnish Business and Policy Forum EVA
Henrik Ehrnrooth, Chairman, Caverion Corporation, Otava and Pöyry PLC
Jorma Ollila, Chairman, Royal Dutch Shell, plc; Chairman, Outokumpu PLC
Risto K. Siilasmaa, Chairman of the Board of Directors and Interim CEO, Nokia Corporation
Kari Stadigh, President and CEO, Sampo PLC
Björn Wahlroos, Chairman, Sampo PLC

France

Henri de Castries, Chairman and CEO, AXA Group
François Baroin, Member of Parliament (UMP); Mayor of Troyes
Nicolas Baverez,Partner: Gibson, Dunn & Crutcher LLP
Pierre-André de Chalendar, Chairman and CEO, Saint-Gobain
Emmanuel Macron, Deputy Secretary General of the Presidency
Natalie Nougayrède, Director and Executive Editor, Le Monde
Fleur Pellerin, State Secretary for Foreign Trade

Germany

Paul M. Achleitner, Chairman of the Supervisory Board, Deutsche Bank AG
Josef Ackermann, Former CEO, Deutsche Bank AG
Jörg Asmussen, State Secretary of Labour and Social Affairs
Mathias Döpfner, CEO, Axel Springer SE
Thomas Enders, CEO, Airbus Group
Norbert Röttgen, Chairman, Foreign Affairs Committee, German Bundestag

Great Britain

Marcus Agius, Non-Executive Chairman, PA Consulting Group
Helen Alexander, Chairman, UBM plc
Edward M. Balls, Shadow Chancellor of the Exchequer
Sherard Cowper-Coles, Senior Adviser to the Group Chairman and Group CEO, HSBC Holdings plc
Robert Dudley, Group Chief Executive, BP plc
Douglas J. Flint, Group Chairman, HSBC Holdings plc
Justine Greening, Secretary of State for International Development
John Kerr, Deputy Chairman, Scottish Power
Peter Mandelson, Chairman, Global Counsel LLP
John Micklethwait, Editor-in-Chief, *The Economist*
George Osborne, Chancellor of the Exchequer GBR Sawers, John Chief, Secret Intelligence Service
Martin H. Wolf, Chief Economics Commentator, *The Financial Times*
John Sawers, Chief, Secret Intelligence Service

Greece

Alexandra Mitsotaki, Chair, ActionAid Hellas
Loukas Tsoukalis, President, Hellenic Foundation for European and Foreign Policy
George Zanias, Chairman of the Board, National Bank of Greece

Hungary

Gordon Bajnai, Former Prime Minister, Party Leader, Together 2014

Ireland
Simon Coveney, Minister for Agriculture, Food and the Marine

Italy
Franco Bernabè, Chairman, FB Group SRL
John Elkann, Chairman, Fiat S.p.A.
Monica Maggioni, Editor-in-Chief, Rainews24, RAI TV
Mario Monti, Senator-for-life; President, Bocconi University

Netherlands
Ben van Beurden, CEO, Royal Dutch Shell plc
Victor Halberstadt, Professor of Economics, Leiden University
Her Royal Highness Princess Beatrix of the Netherlands
Diederik M. Samsom, Parliamentary Leader PvdA (Labor Party)
Paul J. Scheffer, Author; Professor of European Studies, Tilburg University
Edith Schippers, Minister of Health, Welfare and Sport
Gerrit Zalm, Chairman of the Managing Board, ABN-AMRO Bank N.V.

Norway
Svein Richard Brandtzæg, President and CEO, Norsk Hydro ASA
Leif O. Høegh, Chairman, Höegh Autoliners AS
Westye Høegh, Senior Advisor, Höegh Autoliners AS
Eivind Reiten, Chairman, Klaveness Marine Holding AS
Christian Rynning-Tønnesen, President and CEO, Statkraft AS
Jens Ulltveit-Moe, Founder and CEO, Umoe AS

Portugal
Francisco Balsemão, Pinto Chairman, Impresa SGPS
Paulo Macedo, Minister of Health
Inês de Medeiros, Member of Parliament, Socialist Party

Spain
Juan Luis Cebrián, Executive Chairman, Grupo PRISA
Her Majesty the Queen of Spain

Sweden
Carl Bildt, Minister for Foreign Affairs
Håkan Buskhe, President and CEO, Saab AB
Tove Lifvendahl, Political Editor in Chief, *Svenska Dagbladet*
Carl-Henric Svanberg, Chairman, Volvo AB and BP plc
Jacob Wallenberg, Chairman, Investor AB
Marcus Wallenberg, Chairman of the Board of Directors, Skandinaviska Enskilda Banken AB

Switzerland
André Kudelski, Chairman and CEO, Kudelski Group
Daniel L. Vasella, Honorary Chairman, Novartis International

Turkey
Cengiz Çandar, Senior Columnist, *Al Monitor* and *Radikal*
Nilüfer Göle, Professor of Sociology, École des Hautes Études en Sciences Sociales
Mustafa Koç, Chairman, Koç Holding A.S.

Umut Oran, Deputy Chairman, Republican People's Party (CHP)

A. Ümit Taftali, Member of the Board, Suna and Inan Kiraç Foundation

United States

Keith B. Alexander, Former Commander, U.S. Cyber Command; Former Director, National Security Agency

Roger C. Altman, Executive Chairman, Evercore

Nicolas Berggruen, Chairman, Berggruen Institute on GovernanceThomas E. Donilon, Senior Partner, O'Melveny and Myers; Former U.S. National Security Advisor

Martin S. Feldstein, Professor of Economics, Harvard University; President Emeritus, NBER

Michael Gfoeller, Independent Consultant

Evan G. Greenberg, Chairman and CEO, ACE Group

Susan Hockfield, President Emerita, Massachusetts Institute of Technology

Reid Hoffman, Co-Founder and Executive Chairman, LinkedIn

Shirley Ann Jackson, President, Rensselaer Polytechnic Institute

Kenneth M. Jacobs, Chairman and CEO, Lazard

James A. Johnson, Chairman, Johnson Capital Partners

Alex Karp, CEO, Palantir Technologies

Bruce J. Katz, Vice President and Co-Director, Metropolitan Policy Program, The Brookings Institution

Henry A. Kissinger, Chairman, Kissinger Associates, Inc.

Klaus Kleinfeld, Chairman and CEO, Alcoa

Henry R. Kravis, Co-Chairman and Co-CEO, Kohlberg Kravis Roberts & Co.

Marie-Josée Kravis, Senior Fellow and Vice Chair, Hudson Institute

Cheng Li, Director, John L.Thornton China Center, The Brookings Institution

Andrew McAfee, Principal Research Scientist, Massachusetts Institute of Technology

Craig J. Mundie, Senior Advisor to the CEO, Microsoft Corporation

Charles A. Murray, W.H. Brady Scholar, American Enterprise Institute for Public Policy Research

Richard N. Perle, Resident Fellow, American Enterprise Institute

David H. Petraeus, Chairman, KKR Global Institute

Kasim Reed, Mayor of Atlanta

Robert E. Rubin, Co-Chair, Council on Foreign Relations; Former Secretary of the Treasury

Eugene Rumer, Senior Associate and Director, Russia and Eurasia Program, Carnegie Endowment for International Peace

Eric E. Schmidt, Executive Chairman, Google Inc.

Clara Shih, CEO and Founder, Hearsay Social

A. Michael Spence, Professor of Economics, New York University

Lawrence H. Summers, Charles W. Eliot University Professor, Harvard University

Peter A. Thiel, President, Thiel Capital

Kevin M. Warsh, Distinguished Visiting Fellow and Lecturer, Stanford University

James D. Wolfensohn, Chairman and CEO, Wolfensohn and Company

Robert B. Zoellick, Chairman, Board of International Advisors, The Goldman Sachs Group

International

Philip M. Breedlove, Supreme Allied Commander Europe

Benoît Coeuré, Member of the Executive Board, European Central Bank

Christine Lagarde, Managing Director, International Monetary Fund

Anders Fogh Rasmussen, Secretary General, NATO

Viviane Reding, Vice President and Commissioner for Justice, Fundamental Rights and Citizenship, European Commission

Ahmet Üzümcü, Director-General, Organization for the Prohibition of Chemical Weapons

It is believed that some attendees decline to have their names published due to laws such as the 1948 Logan Act (18 U.S.C.A. § 953), which makes it a felony for any member of the federal or state government to meet with members of a foreign government without the authorization of the president or congress.

Notes

INTRODUCTION

Corrupt companies: http://www.huffingtonpost.ca/2013/09/18/world-bank-corrupt-companies-canada_n_3948280.html

Russell Sage Foundation research: http://finance.yahoo.com/blogs/daily-ticker/for-most-families--wealth-has-vanished-172130204.html

Paul Krugman on offshore accounts: http://krugman.blogs.ny-times.com/2014/04/11/offshore-and-underground/?_php=true&_type=blogs&_r=0

Renunciation tax and new legislation: http://www.forbes.com/sites/robertwood/2014/08/28/u-s-hikes-fee-to-renounce-citizenship-by-422/

Forbes list of billionaires: http://www.forbes.com/special-report/2014/billionaires/newcomers.html

Senator Bernie Sanders on ownership of big pizza: http://www.rawstory.com/rs/2014/04/11/bernie-sanders-greedy-billionaires-twisting-american-dream-into-nightmarish-oligarchy/

Will Hutton on Thomas Piketty's economic views: http://www.theguardian.com/commentisfree/2014/apr/12/capitalism-isnt-working-thomas-piketty?CMP=fb_gu

Noam Chomsky on not knowing: http://noam-chomsky.tumblr.com/post/21585364389/the-general-population-doesnt-know-whats

1. DEPOPULATION

Monument as proof of link to New World Order: http://vigilantcitizen.com/sinistersites/sinister-sites-the-georgia-guidestones/

Prince Philip as a virus: http://www.prisonplanet.com/Pages/100604_prince_philip.html

Prince Philip on culling surplus populations: http://modernhistoryproject.org/mhp?Article=CullTheHerd

Maxwell Taylor on population reduction: Editors, "Maxwell Taylor: 'Write Off a Billion,' *Executive Intelligence Review*, September 22, 1981.

NSSM study: http://pdf.usaid.gov/pdf_docs/PCAAB500.pdf

Good Club: http://www.thesundaytimes.co.uk/sto/news/world_news/article169829.ece

Dave Hodges on common foe is the elite: http://www.thecommonsense-show.com/2014/04/26/will-humanity-survive-the-depopulation-agenda-of-the-global-elite/

Dr. Len Horowitz on message is the same: http://www.whale.to/v/scrapie.html

Donald and William Scott on high-level agenda: Donald W. Scott and William L.C. Scott, *The Brucellosis Triangle* (Sudbury, Ontario: the Chelmsford Publishers, 1998), p. 12.

Marian Swain on slowing population growth: http://www.thebreak-through.org/index.php/programs/conservation-and-development/four-surprising-facts-about-population#ixzz2xxdkPv5Y

Justice Ruth Bader Ginsburg on population concern and *Roe v. Wade:* http://www.cnsnews.com/public/content/article.aspx?RsrcID=50819

William Norman Grigg: http://www.lewrockwell.com/grigg/grigg-w102.html

G. Edward Griffin on Holdren's plans for population reduction: http://www.heartcom.org/choice4health.htm

Catherine Austin-Fitts on pandemic as depopulation method: http://www.scoop.co.nz/stories/HL0907/S00250.htm

Oklahoma accommodates 19.49 billion people: www.clover.okstate.edu/fourh/aitc/lessons/upper/sprawl.pdf

Mandy Stahre and birth studies: http://www.cnn.com/2014/03/01/health/cohen-birth-defects/

Nurse Sara Barron: http://www.cnn.com/2014/03/01/health/cohen-birth-defects/

2. THE DEADLY GOD SYNDICATE

Military hardware: http://www.washingtonpost.com/blogs/wonkblog/wp/2013/01/07/everything-chuck-hagel-needs-to-know-about-the-defense-budget-in-charts/

Global arms sales by nation: http://www.globalissues.org/article/74/the-arms-trade-is-big-business#GlobalArmsSalesBySupplierNations

U.S. first in arms transfer agreements: fas.org/sgp/crs/weapons/R42017.pdf

Arms trade corruption: http://www.transparency.org/topic/detail/defence_security

Richard F. Grimmett: https://www.fas.org/sgp/crs/weapons/

Katherine Nightingale: http://www.ipsnews.net/2008/10/politics-unchecked-arms-trade-fuelling-conflict-poverty/

Leland Yee indicted for arms trafficking: http://www.reuters.com/article/2014/04/04/us-usa-california-yee-idUSBREA331K720140404

Direct Commercial Sales: http://lmdefense.com/foreign-military-sales/fms-vs-dcs/

The American Federation of Scientists on lack of DCS oversight: http://fas.org/asmp/library/handbook/WaysandMeans.html

Arms to Syria: http://in.reuters.com/article/2013/06/14/syria-crisis-scenarios-idINDEE95D01620130614

Syrian rebels linked to al-Qaeda: http://www.washingtonexaminer.com/obama-waives-ban-on-arming-terrorists-to-allow-aid-to-syrian-opposition/article/2535885

Hugh Griffiths on well-planned military logistics operation: http://www.nytimes.com/2013/03/25/world/middleeast/arms-airlift-to-syrian-rebels-expands-with-cia-aid.html?pagewanted=all&_r=0

Kevin Shipp, John Rosenthal, and Clare Lopez on secret of the Obama White House: http://www.wnd.com/2015/01/generals-conclude-obama-backed-al-qaida/#eggzyDxDTuy4zjie.99

Chuck Baldwin on perpetual war as number one excuse: http://chuckbaldwinlive.com/Articles/tabid/109/ID/1203/Globalist-Republicans-Pounce-On-Rand-Paul.aspx

ASCE's report card: http://www.infrastructurereportcard.org/

The *Guardian* Statistics: http://www.theguardian.com/news/datablog/2013/sep/17/gun-crime-statistics-by-us-state

Chicago has more shootings than eleven states: http://www.freedomsphoenix.com/Opinion/159289-2014-07-11-the-war-against-our-natural-right-of-self-defense-continues.htm?From=News

Eric Holder on yearly average of mass shootings: http://rt.com/usa/holder-mass-shootings-triple-519/

Mass shootings every sixty-four days since 2011: http://www.huffingtonpost.com/2014/10/15/mass-shootings-tripled_n_5992702.html

Deyfon Pipkin shot during break-in: http://www.myfoxdfw.com/story/21237261/homeowner-kills-intruder-in-oak-cliff-home

Police Chief Gary Hester on Large shooting: http://www.wtsp.com/story/news/crime/2014/04/07/winter-haven-suspect-home-invasion-killed-pena-mitchell-large-wtsp/7435651/

NRA's Chris Cox on gun grab treaty: http://www.usnews.com/opinion/blogs/pat-garofalo/2013/09/26/the-obama-administration-signs-the-un-gun-control-treaty-and-the-nra-freaks-out

Dorothy Stoneman on ending poverty: http://www.huffingtonpost.com/dorothy-stoneman/poverty-gun-violence_b_3528888.html

Oil supply is dwindling: http://usatoday30.usatoday.com/money/economy/2007-07-16-3280171698_x.htm

Peak oil foreseen: Oil supply is dwindling: http://usatoday30.usatoday.com/money/economy/2007-07-16-3280171698_x.htm

William J. Cummings on easy oil and gas found: Stephen Leeb, *Game Over: How You Can Prosper in a Shattered Economy* (New York: Grand Central Publishing, 2009).

CERA 2006 report on Hubbert's methodology: http://alternativeenergy.procon.org/view.answers.php?questionID=001257

IEA on emotive subject of peak oil: http://www.greencarcongress.com/2007/07/iea-sees-oil-su.html

Bakken oil formation: http://www.usgs.gov/newsroom/article.asp?ID=1911#.U3u6PHa9b-E

NRC estimates: http://oils.gpa.unep.org/facts/quantities.htm

Oklahoma Geology Survey on unprecedented earthquakes: http://www.mintpressnews.com/oklahoma-gets-hit-20-earthquakes-one-day-fracking-suspected-cause/195653/

Katie Keranen on events caused by injection: http://www.motherjones.com/environment/2013/03/does-fracking-cause-earthquakes-wastewater-dewatering

Zack Malitz on fighting fracking: http://rt.com/usa/162616-california-senate-kills-fracking-ban/

Credo Action on results of fracking: http://act.credoaction.com/sign/fracking_ab1301

3. DRUGS

WHO Drug Report: http://www.cbsnews.com/news/us-leads-the-world-in-illegal-drug-use/

Ending the Drug War: http://www.lse.ac.uk/IDEAS/Projects/IDPP/The-Expert-Group-on-the-Economics-of-Drug-Policy.aspx

John Collins on drug war failure: http://www.huffingtonpost.com/2014/05/06/end-drug-war_n_5275078.html

David Kupelian on same dead end: http://www.wnd.com/2014/02/the-real-zombie-apocalypse/#VJXDRtSpVbF1ovkc.99

4. DEADLY MEDICINE

One million deaths per decade: t.wordpress.com/2009/12/09/an-exclusive-interview-with-dr-barbara-starfield-medically-caused-death-in-america/

Dr. Barbara Starfield on public hoodwinked: http://jonrappoport.wordpress.com/2009/12/09/an-exclusive-interview-with-dr-barbara-starfield-medically-caused-death-in-america/

CDC on CRE fatality rate: http://www.cdc.gov/hai/organisms/cre/

Dr. Joshua Thaden on CRE case increase: http://www.shea-online.org/View/ArticleId/298/Cases-of-Drug-Resistant-Superbug-Significantly-Rise-in-Southeastern-U-S.aspx

Dr. Joseph Mercola on Japan's understanding: http://articles.mercola.com/sites/articles/archive/2000/07/30/doctors-death-part-one.aspx

Mike Adams on lifestyle management drugs: http://www.naturalnews.com/046041_CRE_superbugs_drug-resistant_infections_modern_plague.html#ixzz37rBaxK5s

Greatest health conspiracy: http://www.naturalnews.com/045075_spontaneous_healing_medical_intervention_health_conspiracies.html#ixzz-31R6nLytV

Dr. Sandy Kweder on taking time: http://www.propublica.org/article/tylenol-mcneil-fda-use-only-as-directed

Representative Walter Jones on ugly roll-call vote: http://www.cbsnews.com/news/under-the-influence/

Cancer costs in 2020: http://www.nih.gov/news/health/jan2011/nci-12.htm

Chemotherapy causes more cancer: http://www.nature.com/nm/journal/v18/n9/full/nm.2890.html

Dave Mihalovic on risks exceed benefits: http://www.wakingtimes.com/2014/03/31/97-percent-time-chemotherapy-work-continues-used-one-reason/

Peter Glidden on chemo profitability: http://preventdisease.com/news/14/033114_97-Percent-of-The-Time-Chemotherapy-Does-Not-Work.shtml; https://www.youtube.com/user/YoungevityEducation

Dr. Cristina Sanchez on THV killing cancer cells: http://naturalsociety.com/molecular-biologist-explains-thc-kills-cancer-completely/#ixzz32ZmdDfGz

Dr. Otto Warburg on prime cause of cancer: Robert N. Proctor, *The Nazi War on Cancer* (Princeton, NJ: Princeton University Press, 1999).

Rita Rubin on serious deficiencies in the FDA: http://usatoday30.usatoday.com/news/health/2004-10-12-vioxx-cover_x.htm

Dr. Joseph Mercola on deaths easily avoided: http://articles.mercola.com/sites/articles/archive/2012/05/14/mercks-adhd-drugs-unsafe.aspx

Criticism of ghostwritten documents: http://medheadlines.com/2008/04/hired-writers-not-scientists-behind-mercks-vioxx-studies/

Martha Rosenberg on hooking public for life: http://www.alternet.org/story/155170/6_kinds_of_pills_big_pharma_tries_to_get_you_hooked_on_for_life?akid=8663.1082462.Cy0eRf&rd=1&t=5

Arianna Huffington on culture of cronyism: http://www.salon.com/2004/11/25/vioxx/

Marc Lipsitch and the studies of mutant viruses: http://www.theguardian.com/world/2014/may/20/virus-experiments-risk-global-pandemic

5. DEADLY VACCINES

Dr. Lawrence Palevsky on standard teaching: http://www.healthychild.com/leading-pediatrician-discusses-vaccines-for-children/

Heartbreaking seizures: http://articles.mercola.com/sites/articles/archive/2009/11/14/Expert-Pediatrician-Exposes-Vaccine-Myths.aspx

Vaccine safety not proven: http://www.nvic.org/NVIC-Vaccine-News/October-2012/Mercola-Palevsky-Interview.aspx

CDC patent on Ebola virus: http://www.google.com/patents/CA2741523A1?cl=en

Dr. Lawrence Palevsky on not making sense: http://articles.mercola.com/sites/articles/archive/2009/11/14/expert-pediatrician-exposes-vaccine-myths.aspx

Dr. Russell Blaylock on herd immunity myth: http://www.vaccination-council.org/2012/02/18/the-deadly-impossibility-of-herd-immunity-through-vaccination-by-dr-russell-blaylock/#sthash.5EqulFLL.dpuf

Palevsky on informed choice: http://articles.mercola.com/sites/articles/archive/2009/11/14/Expert-Pediatrician-Exposes-Vaccine-Myths.aspx

Dr. Joseph Mercola on increasing diseases: http://articles.mercola.com/sites/articles/archive/2010/11/04/big-profits-linked-to-vaccine-mandates.aspx

Dr. Mary Anne Jackson on D68 in midwest children: http://www.cnn.com/2014/09/06/health/respiratory-virus-midwest/

June Valent wins lawsuit: http://www.newsmaxhealth.com/Health-News/flu-shot-nurse-fired/2014/06/05/id/575407/

GlaxoSmithKline releases polio virus in Belgium: http://vaccinenewsdaily.com/medical_countermeasures/331649-ecdc-provides-assessment-of-accidental-polio-incident-in-belgium/

Dr. Perry Kendall on Canadian study causes disarray: http://www.theglobeandmail.com/technology/science/more-flu-programs-suspended/article4287080/

WHO's Margaret Chan on catastrophic loss of lives: http://www.bbc.com/
news/world-africa-28610112

Liberian community care centers: http://www.washingtonpost.com/
national/health-science/new-effort-to-fight-ebola-in-liberia-would-
move-infected-patients-out-of-their-homes/2014/09/22/f869dc08-4281-
11e4-b47c-f5889e061e5f_story.html

State Department bids for 160,000 hazmat suits: http://finance.yahoo.com/
news/lakeland-industries-announces-global-availability-142200024.html

Mike Adams on medical theater: http://www.naturalnews.com/046290_
Ebola_patent_vaccines_profit_motive.html#ixzz39MgOLJpg

Kurt Nimmo on manufactured crises: http://www.prisonplanet.com/dont-
fear-ebola-fear-the-state.html

Dr. Eric Pianka and Forrest Mims on population reduction by Ebola: http://
www.pearceyreport.com/archives/2006/04/transcript_dr_d.php/

Options limited to one system: http://www.naturalnews.com/044879_
MERS_pandemic_Saudi_Arabia_drug-resistant_superbugs.html

Dr. Joseph Mercola on thimerosal in low concentrations: http://articles.
mercola.com/sites/articles/archive/2009/08/06/Proof-That-Thimerosal-
Induces-AutismLike-Neurotoxicity.aspx

Boyd Haley on injection of thimerosal: http://www.vaccinesuncensored.
org/ingredients.php

New policy of not paying doctors: http://online.wsj.com/news/articles/SB1
0001424052702303532704579476862300832806

Eliot Spitzer on charging individual CEOs: http://www.nytimes.
com/2012/07/03/business/glaxosmithkline-agrees-to-pay-3-billion-in-
fraud-settlement.html?pagewanted=all

Ivan Oransky on rarely held to account: http://abcnews.go.com/Technol-
ogy/wireStory/researcher-charged-major-hiv-vaccine-fraud-case-24288252

Dr. William Thompson on CDC fraud concerning MMR and autism:
http://ireport.cnn.com/docs/DOC-1164046

William W. Thompson on I have stopped lying: http://jonrappoport.word-
press.com/

William W. Thompson on omission of significant information: http://www.
morganverkamp.com/august-27-2014-press-release-statement-of-william-
w-thompson-ph-d-regarding-the-2004-article-examining-the-possibility-
of-a-relationship-between-mmr-vaccine-and-autism/

Stephen Krahling and Joan Wlochowski lawsuit against Merck: http://
jonrappoport.wordpress.com/2014/09/24/there-are-2-other-mmr-
vaccine-whistleblowers/;   http://www.plainsite.org/dockets/mmpl79tu/
pennsylvania-eastern-district-court/united-states-of-america-et-al-v-merck-
and-co/

Audrey Rzhetsky on tiny molecules: http://www.naturalnews.
com/044649_mercury_pollution_autism_flu_shots.html

Dr. Audrey Rzhetsky: http://news.uchicago.edu/article/2014/03/13/
autism-and-intellectual-disability-incidence-linked-environmental-
factors#sthash.AaIIeIg2.dpuf

Mercury as hazardous material: http://www.naturalnews.com/011764_thi-
merosal_mercury.html#ixzz2q8f5pVD2

Scientific proof needed: Dr. Lawrence Palevsky, "Aluminum and Vaccine

Ingredients: What Do We Know and What Don't We Know," National Vaccine Information Center, Doctor's Corner, March 2008.

Representative Dan Burton on absolute disaster: http://healthimpactnews. com/2012/video-highlights-from-first-congressional-hearing-on-autism-in-10-years/

Burton on research needed now: http://web.archive.org/ web/20060316141931/http://ccmadoctors.ca/opening_statement_chairman_dan_b1.htm

Hooker on globalization of vaccines: http://www.ageofautism. com/2012/12/brian-hookers-testimony-autism.html

Twenty-six studies in other countries: http://articles.mercola.com/sites/articles/archive/2010/04/10/wakefield-interview.aspx

Dr. David Lewis on no fraud: http://www.prnewswire.com/news-releases/ ongoing-investigations-by-dr-david-lewis-refute-fraud-findings-in-dr-andrew-wakefield-case-133649563.html

Dr. Lewis on research manipulation: http://www.whistleblowers.org/index. php?option=com_content&task=view&id=1220&Itemid=190

Dr. Thomas Verstraeten and FOIA document: http://www.autisminvestigated.com/tag/dr-thomas-verstraeten/

Fake vaccination program endangers us all: http://www.scientificamerican. com/article/how-cia-fake-vaccination-campaign-endangers-us-all/

Cancer-causing virus in polio vaccine: http://www.naturalnews. com/045318_fake_vaccines_DNA_harvesting_White_House.html#ixzz-333JcUAq1

VICP funding: http://www.hrsa.gov/vaccinecompensation/index.html

Judge Sotomayor on regulatory vacuum: http://www.csmonitor.com/USA/ Justice/2011/0222/Parents-can-t-sue-drug-firms-when-vaccines-cause-harm-Supreme-Court-says

Couric's misdeeds: http://ideas.time.com/2013/12/04/is-katie-couric-the-next-jenny-mccarthy/

Katie Couric's personal view: http://www.huffingtonpost.com/katie-couric/vaccine-hpv-furthering-conversation_b_4418568.html

Barbara Loe Fisher on well-orchestrated campaign: http://www.nvic.org/ NVIC-Vaccine-News/January-2014/Gardasil-Vaccine-and-Katie-Couric. aspx

Vaccine annual growth rate: Top 5 Vaccine Companies by Revenue—2012—FierceVaccines http://www.fiercevaccines.com/special-reports/top-5-vaccine-companies-revenue-2012#ixzz30Dx3E9DJ

Amy Wallace: http://www.wired.com/2009/10/ff_waronscience/

Mike Adams on "antiscience": http://www.naturalnews.com/z044620_ Chilis_autism_awareness_week_medical_mafia.html

FluLaval safety not established: http://www.naturalnews.com/045418_ flu_shots_influenza_vaccines_mercury.html#ixzz33avH19n4

Mumps outbreak: http://www.nejm.org/doi/full/10.1056/NEJ-Moa1202865#t=articleTop

6. GONE TO POT

2013 marijuana survey: http://www.usatoday.com/story/news/ nation/2013/09/04/marijuana-drug-use-survey/2760061/

Alfredo Carrasquillo on pot arrests in New York City: http://www.huffingtonpost.com/2013/03/19/nypd-marijuana-arrests_n_2908285.html

Michigan couple loses daughter: http://jonathanturley.org/2013/09/24/michigan-child-services-remove-baby-from-home-due-to-parents-use-of-legal-medical-marijuana/

Kelley Vlahos on pot money going into system: http://www.unz.com/article/war-on-drugs-ends-with-a-fizzle/

Jordan Cornelius and Mac Clouse on feds extending the drug war: http://www.usatoday.com/story/news/nation/2014/11/03/irs-limits-profits-marijuana-businesses/18165033/

ACLU study on hyperaggressive domestic policing: https://www.aclu.org/criminal-law-reform/war-comes-home-excessive-militarization-american-police-report

Jared Polis and Dana Rohrabacher on pro-pot amendment: http://www.huffingtonpost.com/2014/05/30/medical-marijuana-congress_n_5418084.html

Michele Leonhart on fighting harder: http://www.huffingtonpost.com/2014/04/02/dea-marijuana-legalization_n_5078046.html?1396465677

Pew Research Center survey of drugs: http://www.people-press.org/2014/04/02/americas-new-drug-policy-landscape/

Steph Sherer and Darryl Moore on federal suit: http://www.huffingtonpost.com/2013/05/08/california-medical-marijuana-crackdown_n_3241324.html

Lisa Eubanks on THC use in stopping Alzheimer's: http://naturalsociety.com/4-solutions-preserving-brain-health-preventing-alzheimers/#ixzz2valGQ

Sayer Ji on more American prisoners than old Soviet Union: http://www.wakingtimes.com/2014/09/02/cannnabis-future-medicine/

Michael Boldin on hemp growing despite federal ban: http://benswann.com/federal-government-legalizes-hemp/

Mitch McConnell, Rand Paul, and impounded hemp seeds: http://www.naturalnews.com/045285_hemp_seeds_Kentucky_DEA.html

Adam Watson on growing hemp: http://abcnews.go.com/US/wireStory/kentucky-gains-permit-hemp-seeds-23834018?singlePage=true

Dan DuBray, Ellen Ca+nale, and the eight factors: http://www.nbcnews.com/news/us-news/u-s-says-legal-marijuana-growers-cant-use-federal-irrigation-n110381

Elan Nelson on federal obstacles: http://www.nbcnews.com/news/us-news/u-s-says-legal-marijuana-growers-cant-use-federal-irrigation-n110381

Cristie Drumm on not banking marijuana businesses: http://www.denverpost.com/marijuana/ci_25180402/colorado-governor-reveals-pot-tax-spending-plan

Rachel Gillette on money laundering: http://www.wnd.com/2014/07/dont-try-paying-irs-in-cash/#7uKZW14ljUqpItA3.99

Angela Kirking and DEA raid: http://www.huffingtonpost.com/2014/04/12/dea-marijuana-garden-store_n_5128771.html?ncid=txtlnkusaolp00000592

Ben Siller and his olfactometer: http://www.npr.org/blogs/thetwo-way/2013/11/14/245254291/denvers-smell-o-scope-targets-marijuanas-skunky-scent

Mike Sledge and Harte raid: http://www.huffingtonpost.com/2014/04/12/dea-marijuana-garden-store_n_5128771.html?ncid=tx-tlnkusaolp00000592

Joe Lavoro and Jack Holmes on pot sentence: http://kfor.com/2014/05/16/im-frightened-for-my-son-teen-faces-life-in-prison-for-pot-brownies/

Pot seizures: http://fusion.net/justice/story/analysis-drug-porn-exciting-anymore-15359

Prescription drugs cause more deaths than illegal drugs: http://www.drug-freeworld.org/drugfacts/prescription/abuse-international-statistics.html

WHO survey: http://www.cbsnews.com/news/us-leads-the-world-in-illegal-drug-use/

President José Mujica legalizing pot in Uruguay: http://www.rawstory.com/rs/2014/03/12/uruguay-planning-to-sell-legally-cloned-marijuana-by-years-end/

Mason Tvert on tax revenues: http://bigstory.ap.org/article/colorado-governor-reveals-pot-tax-spending-plan

Reports on pot from Denver: http://www.rawstory.com/rs/2014/07/03/six-months-after-marijuana-legalization-colorado-tax-revenue-skyrockets-as-crime-falls/

Lee Fang on painkiller consultants opposing legal marijuana: http://www.vice.com/read/leading-anti-marijuana-academics-are-paid-by-painkiller-drug-companies

7. PSYCHIATRIC DRUGS AND SHOOTERS

Bianca de Kock on smirky smile: http://radaronline.com/exclu-sives/2014/05/elliot-rodger-smile-murder-killings-evil-bianca-de-kock/

Elliot Rodger in his own words: http://beforeitsnews.com/alterna-tive/2014/05/the-real-reason-elliot-rodger-opened-fire-on-innocent-victims-who-would-have-guessed-disturbing-video-2965188.html

Roberto Malinow on playing with memory: http://www.nature.com/news/flashes-of-light-show-how-memories-are-made-1.15330

Dr. David Healy on emotional numbness: http://prorevnews.blogspot.com/2013/09/mass-murders-and-antidepressants.html

Sam Smith on twenty-five thousand mass killers: http://prorevnews.blog-spot.com/2013/09/mass-murders-and-antidepressants.html

8. DRUGGING THE MILITARY:

Jessie Jane Duff on mismanagement at the VA: www.wnd.com/2014/06/marine-watchdog-va-deaths-actually-in-thousands/#DQbMEmscxbk9CSZS.99

John Dickerson on growing VA scandal: http://www.slate.com/articles/news_and_politics/politics/2014/05/veterans_affairs_scandal_why_the_treatment_of_our_veterans_is_a_genuine.html

Pauline DeWenter on bringing vets back to life: http://www.cnn.com/2014/06/23/us/phoenix-va-deaths-new-allegations/

John W. Whitehead on lack of protection by government: http://www.le-wrockwell.com/2014/05/john-w-whitehead/the-state-is-afraid-of-its-own-veterans/

Ron Paul on most egregious abuse of armed forces: http://ronpaulinstitute.org/archives/featured-articles/2014/may/25/the-va-scandal-is-just-the-tip-of-the-military-abuse-iceberg.aspx

Representative Jack Metcalf irritated by squalene assurances: http://www.dynamicchiropractic.com/mpacms/dc/article.php?id=31995

Captain Richard Rovet on servicemen and women as test subjects: http://www.naturalnews.com/042241_gulf_war_syndrome_flu_vaccines_mf59.html#ixzz3Lpn6GVpG

9. DEADLY FOOD

Sandwich study on processed food sandwich: http://www.foodandnutri-tionresearch.net/index.php/fnr/article/view/5144

Kris Gunnar on processed food addiction, whole grains, and real oils: http://authoritynutrition.com/9-ways-that-processed-foods-are-killing-people/

Dr. Yaneer Bar-Yam on corn prices causing unrest: http://dailysignal.com/2013/02/07/ethanol-mandate-leads-to-social-unrest/

Feces in food: http://www.infowars.com/there-is-a-staggering-amount-of-feces-in-our-food/

Contaminated seafood: http://www.bloomberg.com/news/2012-10-11/asian-seafood-raised-on-pig-feces-approved-for-u-s-consumers.html

Steve Crider on growth of organics: http://www.politico.com/story/2014/05/organic-food-industry-civil-war-agriculture-usda-106295.html#ixzz30xLfrDif

NOSB allows nonorganic materials: http://www.organicconsumers.org/articles/article_29653.cfm

Alexis Baden-Mayer and Ronnie Cummins on unlikely board vote: http://www.organicconsumers.org/articles/article_29653.cfm

Poll shows ignorance of antibiotics in fruit: http://www.takepart.com/article/2013/04/06/antibiotic-use-organic-apples-pears

Mineral loss in wheat: http://www.ncbi.nlm.nih.gov/pubmed/19013359

10. DEADLY GRAINS

Adulterated bread: http://www.naturalnews.com/043499_glyphosate_contamination_cereal_bars.html#ixzz2qJC22bTy

USDA does not test for glyphosate: http://www.reuters.com/article/2014/12/19/us-usda-pesticides-report-idUSKBN0JX2FZ20141219

Glyphosate: http://www.naturalnews.com/041464_glyphosate_Mon-santo_toxicity.html#ixzz2qJFm4Bns

Dr. Stephanie Seneff on glyphosate and autism: http://www.anh-usa.org/half-of-all-children-will-be-autistic-by-2025-warns-senior-research-scientist-at-mit/

2014 study of Roundup toxicity: http://www.ncbi.nlm.nih.gov/pmc/articles/PMC3955666/

Sayer Ji on no longer an acceptable level of harm: http://www.greenmedinfo.com/blog/roundup-herbicide-125-times-more-toxic-regulators-say?page=2

Bromide treats psychosis: http://ajp.psychiatryonline.org/article.aspx?articleID=143181

Bromide in thyroid cancer: http://breastcancerchoices.org/bromidedominancetheory.html

Dr. Gary E. Abraham: http://findpdf.net/reader/The-Safe-and-Effective-Implementation-of-Orthoiodosupplementation.html

Dr. Lawrence Wilson: http://drlwilson.com/ARTICLES/BREAD.htm

Reporters fired: http://intellihub.com/2013/06/15/fox-reporters-fired-for-reporting-the-truth-about-monsanto-milk/

Vani Hari and petition: http://foodbabe.com/subway/

EWG gets results: http://www.ewg.org/enviroblog/2014/03/kudos-companies-pledging-remove-ada

Dartmouth Medical School researchers on higher levels of arsenic: http://www.webmd.com/diet/features/arsenic-food-faq

Uncle Ben's Rice recall: http://www.fda.gov/NewsEvents/Newsroom/PressAnnouncements/ucm385299.htm

No country of origin statement: http://www.naturalnews.com/044744_rice_protein_heavy_metals_petition_organic_superfoods.html#ixzz2z4TcYjCe

Kris Gunnars on ineffective diet: http://authoritynutrition.com/11-graphs-that-show-what-is-wrong-with-modern-diet/

11. DEADLY SWEETENERS

Sean O'Donnell on the faster they die: http://www.cbsnews.com/news/truvia-sweetener-is-toxic-to-fruit-flies-study-finds/

A novel pest control: http://www.plosone.org/article/info%3Adoi%2F10.1371%2Fjournal.pone.0098949

Mike Adams on mind games and an interesting study: http://www.naturalnews.com/045450_Truvia_erythritol_natural_pesticide.html#ixzz33zZOFXe9

Shane Ellison: http://thepeopleschemist.com/splenda-the-artificial-sweetener-that-explodes-internally/

Dr. Morando Soffritti on leukemia in mice: http://www.medicalnewstoday.com/articles/262475.php

Maureen Conway on safety of Splenda: http://www.splendatruth.com/news

2010 study on links between artificial sweeteners and weight gain: http://www.livestrong.com/article/447584-what-are-the-dangers-of-splenda-sucralose-and-aspartame/

Dr. Louis J. Elsas on high levels of phenylalanine: www.wnho.net/dr_elsas_testimony.pdf

Dr. Joseph Mercola on most dangerous substance: http://articles.mercola.com/sites/articles/archive/2011/11/06/aspartame-most-dangerous-substance-added-to-food.aspx

Aspartame turns to formaldehyde: http://www.mpwhi.com/formaldehyde_from_aspartame.

Zachary Shahan on two-thirds of U.S. population: http://eatdrinkbetter.com/2009/11/15/can-diet-coke-kill-you-part-2/

Dr. Janet Starr Hull and academics: http://www.janethull.com/newsletter/0206/bella_italia_the_soffritti_aspartame_study.php

University of Iowa diet-drink study: http://now.uiowa.edu/2014/03/ui-study-finds-diet-drinks-associated-heart-trouble-older-women

Terri LaPoint on aspartame as a neurotoxin: http://www.inquisitr.com/1194961/new-study-just-two-diet-sodas-a-day-increase-heart-disease-risk-for-women/#4T1K6OIL82iCZJIl.99

Soft-drink sales down: http://time.com/44282/soda-sales-drop/?utm_source=feedburner&utm_medium=feed&utm_campaign=Feed%3A+-time%2Ftopstories+%28TIME%3A+Top+Stories%29

Drs. James Bowen and Betty Martini: http://www.wnho.net/aspartameandgulfwar.htm

Peter Piper on DNA damage: http://www.prevention.com/food/healthy-eating-tips/diet-soda-bad-you/cell-damage

FDA sees no evidence of dioxins: http://dashingclaire.hubpages.com/hub/Dr-Edward-Fujimoto-is-Real-The-Cancer-Update-Emails-Are-Not

University of Calgary study of BPS: http://www.washingtonpost.com/national/health-science/bpa-alternative-disrupts-normal-brain-cell-growth-is-tied-to-hyperactivity-study-says/2015/01/12/a9ecc37e-9a7e-11e4-a7ee-526210d665b4_story.html

12. GMOS:

Dr. Paul Craig Roberts on Dick Chaney: http://www.paulcraigroberts.org

Jeffrey M. Smith on bad science: http://www.globalresearch.ca/monsanto-the-tpp-and-global-food-dominance/5359491

Twenty-seven million dollars to defeat GMO labeling: http://www.huffingtonpost.com/2014/09/03/gmo-labeling-foes_n_5756710.html

Dr. David Bronner on GMO movement: http://www.huffingtonpost.com/david-bronner/dr-bronners-yearend-repor_b_6357178.html

Gary Hooser on GMO ordinance overturned: http://online.wsj.com/articles/u-s-judge-overturns-gmo-crop-curbs-in-hawaii-1409009260

AAEM recommends avoiding GMO feed: http://www.responsibletechnology.org/gmo-dangers

IRT concerns over antibiotic genes: http://www.responsibletechnology.org/gmo-dangers

Non-GMO Project on novel organism cannot be recalled: http://www.nongmoproject.org/learn-more/

Jay Feldman on fatally flawed technology: http://www.mintpressnews.com/environmentalists-rally-new-herbicide-ge-crops/195697/

Thomas J. Goreau on soil carbon: http://e360.yale.edu/feature/soil_as_carbon_storehouse_new_weapon_in_climate_fight/2744/

Rattan Lal on soil sequestration: http://e360.yale.edu/feature/soil_as_carbon_storehouse_new_weapon_in_climate_fight/2744/

Sean Poulter on coup for GM industry: http://www.dailymail.co.uk/news/article-2621058/Secret-emails-reveal-ministers-plotted-GM-lobbyists-Documents-two-worked-campaign-win-sceptical-consumers.html#ixzz3OkEtcsSE

Cannabis to restore soil: http://www.naturalnews.com/045106_GMOs_chemical_agriculture_carbon_emissions.html#ixzz31cj58XTD

Chensheng Lu on honeybee colony collapse: http://www.theguardian.com/environment/2014/may/09/honeybees-dying-insecticide-harvard-study

Sheldon Krimsky questions agribusiness testing of GMOs: http://www.bostonglobe.com/lifestyle/health-wellness/2014/08/17/with-sheldon-krimsky/oR1rIk3yspnUcJvHKtrrbM/story.html

Dr. Helen Wallace on government-GMO collusion: http://www.dailymail.co.uk/news/article-2621058/Secret-emails-reveal-ministers-plotted-GM-lobbyists-Documents-two-worked-campaign-win-sceptical-consumers.html#ixzz34GgdcMw4

Marion Nestle on GMO companies' fear: http://www.cbsnews.com/news/figuring-out-whats-in-your-food/

Andy Kimbrell on tracing food problems: http://www.cbsnews.com/news/figuring-out-whats-in-your-food/

Jerry Greenfield on labeling GMOs not scary: http://www.huffingtonpost.com/2014/07/10/gmo-labels-congress_n_5576255.html

Oxfam on revenues and food-company chart: http://firstperson.oxfamamerica.org/2013/03/10-everyday-food-brands-and-the-few-giant-companies-that-own-them/

Rory Hall on controlled plan working against humanity: http://www.thedailysheeple.com/control-the-food-control-the-people_092014#sthash.N0vxDE7n.dpuf

Tracie McMillan on food price going to farmers: http://eatocracy.cnn.com/2012/08/08/where-does-your-grocery-money-go-mostly-not-to-the-farmer/

13. DEADLY WATER

Miles O'Brien on chromium-6: http://www.pbs.org/newshour/extra/daily_videos/chromium-6-taints-water-supplies-around-the-country/

Jeffrey Kluger and John Spatz on drugs in water: http://content.time.com/time/specials/packages/article/0,28804,1976909_1976907_1976871,00.html

LEEP on phosphorus in Lake Erie: http://www.ijc.org/en_/leep/Home#sthash.uePyDGDv.dpuf

Mike Adams on system declaring life sacred: http://www.naturalnews.com/046314_Toledo_water_crisis_algal_bloom.html#ixzz39RzKKlDe

Olivia Sanchez and lack of water in San Joaquin Valley: http://www.ktvu.com/news/news/local/drought-leaves-california-homes-without-water/ng7yc/

Richard Howitt on slow-moving train wreck: http://www.washingtonpost.com/national/health-science/wests-historic-drought-stokes-fears-of-water-crisis/2014/08/17/d5c84934-240c-11e4-958c-268a320a60ce_story.html

Andrew Liveris on water as twenty-first-century oil: http://www.economist.com/node/11966993

Gary Harrington on being fined and jailed over water collection: http://www.huffingtonpost.com/2012/08/16/gary-harrington-oregon-water-rainwater_n_1784378.html

Jo-Shing Yang on banks and billionaires controlling the water supply: http://mynaturesmedicine.com/2014/05/26/the-new-water-barons-wall-street-mega-banks-are-buying-up-the-worlds-water/

Jim Henson on public apathy about water: http://www.texastribune.org/2013/03/07/uttt-poll-water-priority-not-top-problem/

Dr. Donald Miller on poorly designed fluoride studies: http://archive.le-wrockwell.com/miller/miller17.html

Fluoride study on urinary stones: http://link.springer.com/article/10.1134/S1061934814050128

Regina Imburgia on health risks: http://www.prweb.com/releases/2014/02/prweb11620652.htm

Fluoride opponents founded in fear: http://www.dallasobserver.com/2014-05-08/news/in-dallas-an-anti-flouride-movement-for-once-not-dismissed/full/

York review of studies: http://www.york.ac.uk/inst/crd/fluoridnew.htm

NRC fluoride report of 2006: http://www.nap.edu/catalog.php?record_id=11571

Studies ignored or downplayed: Paul Connett et al, *The Case Against Fluoride: How Hazardous Waste Ended Up in Our Drinking Water and the Bad Science and Powerful Politics That Keep It There* (White River Junction, VT: Chelsea Green Publishing, 2010), p. 86.

Dr. Paul Cornett to Dallas City Council: https://www.youtube.com/watch?v=EBdAF1gqANs&feature=youtu.be

Paul Connett on fluoride strategy: http://healthfreedoms.org/2014/03/18/10-facts-about-fluoride-for-the-skeptic/#sthash.OXtFp5ro.dpuf

Fluoride and free radicals: http://www.nap.edu/openbook.php?record_id=11571&page=222

Antifluoride groups reenergized: Mike Stobbe, "U.S. Plans to Lower Recommended Fluoride Levels," *Associated Press* (January 8, 2011).

Study showed damage to liver and kidney: http://www.ncbi.nlm.nih.gov/pubmed/16834990

2012 joint study on cognitive development: http://www.hsph.harvard.edu/news/features/fluoride-childrens-health-grandjean-choi/

Connett and coauthors on system to perpetuate fluoridation: http://www.whale.to/a/connett_b.html

Thyroid article from 1958: http://www.slweb.org/galletti.html

Dr. Richard L. Shames and his wife, Karilee, on thyroid damage from fluoride: http://thyroid.about.com/od/drsrichkarileeshames/a/fluoridechange.htm

14. DEADLY AIR

Dr. Carlos Dora, WHO, and World Bank studies: http://www.nytimes.com/2014/03/26/world/pollution-killed-7-million-people-worldwide-in-2012-report-finds.html

Steven J. Davis on plenty of blame: http://news.uci.edu/press-releases/made-in-china-for-us-air-pollution-as-well-as-exports/

EPA's Gina McCarthy on kids using inhalers: http://www.cnsnews.com/news/article/susan-jones/mccarthy-if-your-kid-doesnt-use-inhalerconsider-yourself-very-lucky-parent

Fukushima radiation: http://rt.com/news/fukushima-radiation-8-times-standard-448/

Aoyama Otaki and Dr. Timothy Mousseau on subtle censorship: http://www.naturalnews.com/044414_insidious_censorship_radiation_Japan.html

Akira Ono on radioactive water: http://www.japantimes.co.jp/news/2014/04/20/national/fukushima-no-1-boss-admits-water-woes-out-of-control/#.U1VOeqK9b-E

Ken Buesseler on making predictions and models: http://www.usatoday.com/story/news/nation/2014/03/09/scientists-test-west-coast-for-fukushima-radiation/6213849/

Dr. Jeff Patterson on no safe level: http://www.psr.org/news-events/press-releases/psr-concerned-about-reports-increased-radioactivity-food-supply.html

Ivan Macfadyen and a dead ocean: http://www.theherald.com.au/story/1848433/the-ocean-is-broken/

David Hirsch on no fear on the beach: http://www.ksbw.com/news/central-california/santa-cruz/zero-threat-of-fukushima-radiation-at-california-beaches-health-officials-say/23872086

Dana Durnford on missing species: https://www.youtube.com/watch?v=QoBGBYDjxQc

Dr. Timothy A. Mousseau on radiation damage to entire ecosystem: http://enenews.com/scientist-holds-press-conference-tokyo-urgent-need-share-new-findings-fukushima-very-very-striking-results-show-injury-ecosystems-radiation-significant-implications-regions-japan-video

Dr. Shegira Mita on Tokyo radiation victims: http://www.save-children-from-radiation.org/2013/11/11/title-dr-shigeru-mita-addresses-the-need-of-blood-examination-among-children-in-the-kanto-area/

Christina Consolo on airline radiation: http://exopolitics.blogs.com/peaceinspace/2014/02/radchick-fukushima-triggers-unprecedented-increase-in-airline-pilot-passenger-heart-attacks-cancers-radiation-illness-s.html

Wendy Hopkins on naturally occurring materials: http://sanbruno.patch.com/groups/around-town/p/san-mateo-county-beach-radiation-not-from-fukushima-officials-say_de592463

Steven Manley and Kai Vetter head team seeking radiation: http://www.dailycal.org/2014/05/12/sign-yet-fukushima-radiation-us-west-coast-shoreline/

Stephen H. Hanauer and Joseph Hendrie on safety of Mark 1: http://www.nytimes.com/2011/03/16/world/asia/16contain.html

Chris Carrington on Jeffrey Immelt's knowledge of warnings: http://www.globalresearch.ca/why-the-obama-administration-will-not-admit-that-fukushima-radiation-is-poisoning-americans/5365626

Governments created system to protect companies: http://www.greenpeace.org/africa/en/News/news/Fukushima-Fallout/

Naoto Kan on elimination of nuclear power plants: http://www.democracynow.org/2014/3/11/ex_japanese_pm_on_how_fukushima#

NFS uranium leak: http://www.knoxnews.com/news/2007/jul/11/erwin-uranium-spill-cloaked-in-secrecy/

Radiation leak at Los Alamos: http://www.sfgate.com/news/texas/article/New-Mexico-says-57-nuke-containers-could-be-threat-5490447.php

Radioactive water leak in South Carolina: http://www.thestate.com/2014/07/15/3565904/leak-sparks-shutdown-of-atomic.html

Sabotage at Belgian nuclear plant: http://uk.reuters.com/article/2014/08/14/belgium-nuclear-doel-idUKL6N0QK43R20140814

Isabelle Baldi link between cell-phone use and tumors: http://www.theguardian.com/society/2014/may/13/intensive-mobile-phone-users-higher-risk-brain-cancer-study

Dr. Richard A. Stein on cell-phone dangers: http://www.lfpress.com/comment/2010/07/02/14587546.html

Mike Adams on confused and irritable public: http://www.naturalnews.com/z044464_cell_towers_EMF_pollution_mental_confusion.html

Cell-tower hazards: http://bmjopen.bmj.com/content/3/12/e003836.full

Dr. Ilya Sandra Perlingieri on aerosol assault: http://www.globalresearch.ca/chemtrails-the-consequences-of-toxic-metals-and-chemical-aerosols-on-human-health/19047

Carole Pellatt on well-documented military testing on civilians: http://arizonaskywatch.com/articles/articles/Yes,%20We%20Are%20Being%20Sprayed.htm

David Oglesby on grid pattern: http://www.rense.com/general15/chemusmilitarycontinues.htm

Bob Fitrakis on questioning Kucinich: http://www.chemtrailcentral.com/ubb/Forum1/HTML/001113.html

Colonel Walter M. Washabaugh on Chemtrail "hoax": http://www.carnicominstitute.org/articles/af3.htm

Mark Steadham and software: http://www.chemtrailcentral.com/report.shtml

Vaughan Rees on secondhand smoke particles: http://archive.sph.harvard.edu/press-releases/2006-releases/press10052006.html

Therese Aigner on controlled delivery of Chemtrails: http://www.rense.com/general21/conf.htm

Michael J. Murphy on aluminum-resistant Monsanto seeds: http://chemtrailsnorthnz.wordpress.com/opinions-regarding-the-functions-of-chemtrailsstratospheric-aerosol-geoengineering/

Dr. R. Michael Castle on serious skin lesions: http://chemtrailsplanet.net/2012/12/26/3598/

Dr. Russell Blaylock on major concern over nanosized aluminum: http://chemtrailsplanet.net/2013/06/20/aluminum-in-chemtrails-linked-to-rapid-increase-in-neurogenic-disorders/

Foreign Affairs article on geoengineering: http://www.foreignaffairs.com/articles/139084/david-g-victor-m-granger-morgan-jay-apt-john-steinbruner-kathari/the-truth-about-geoengineering

John Holdren on shooting pollutants into the atmosphere: http://www.nypost.com/p/news/politics/bam_man_cool_idea_block_sun_2Opipfl-ho393Yi7gYoJLXP

David Walker on work completed: http://rt.com/usa/167172-haarp-future-uncertain-air-force/

15. A POLICE STATE

John W. Whitehead on incomprehensible police state: https://www.rutherford.org/publications_resources/john_whiteheads_commentary/the_absurd_bureaucratic_hell_that_is_the_american_police_state

Chris Hedges on creating neofeudalism: http://www.truth-out.org/opin-ion/item/21335-chris-hedges-what-obama-really-meant-was

Faisal Gill on shaping different policy: http://www.mintpressnews.com/nsa-bombshell-agency-spied-prominent-american-citizens/193692/

Edward Snowden statement: http://wikileaks.org/Statement-by-Edward-Snowden-to.html

Christopher Ketchum on Main Core list of enemies: http://www.infowars.com/main-core-a-list-of-millions-of-americans-that-will-be-subject-to-detention-during-martial-law/\

Tom Shorrock on Main Core intelligence data: http://www.salon.com/2008/07/23/new_churchcomm/

Kirsten Weld on RTRG program: http://america.aljazeera.com/opin-ions/2014/5/institutionalizingsurveillanceusnsaguatemalaarchive.html

Edward Snowden on passing around nude photographs: http://www.forbes.com/sites/kashmirhill/2014/07/17/nsa-responds-to-snowden-claim-that-intercepted-nude-pics-routinely-passed-around-by-employees/

Chris Hedges on surveillance state: http://www.truthdig.com/report/item/what_obama_really_meant_was_20140119

James Clapper on FISA extension: http://www.theguardian.com/world/2014/jun/21/fisa-court-nsa-collection-metadata

Randal Milch on changes to NSA data collection: http://www.reuters.com/article/2014/04/03/us-usa-security-obama-idUSBREA3228O20140403

William Binney: http://www.theguardian.com/commentisfree/2014/jul/11/the-ultimate-goal-of-the-nsa-is-total-population-control

Hector Garza on easily manipulated paperwork: http://www.breitbart.com/Breitbart-Texas/2014/07/11/Exclusive-TSA-Allowing-Illegals-to-Fly-Without-Verifiable-ID-Says-Border-Patrol-Union

David Jancik and license-plate recognition: http://www.couriermail.com.au/news/opinion/opinion-automatic-number-plate-recognition-allows-police-to-track-drivers8217-movements-further-eroding-our-civil-liberties/story-fnihsr9v-1226719544697

James Smith on abuse of license recognition: http://prepperpodcast.com/warning-homeland-security-set-to-purchase-license-plate-tracking/

Al Franken on abuse of NGI system: http://www.commondreams.org/news/2014/09/16/big-brother-30-fbi-launches-facial-recognition-program

Alessandro Acquisti and Al Franken: http://www.nytimes.com/2014/06/01/us/nsa-collecting-millions-of-faces-from-web-images.html

Jennifer Lynch and MorphoTrust: http://arstechnica.com/tech-policy/2014/04/fbi-to-have-52-million-photos-in-its-ngi-face-recognition-database-by-next-year/

Jay Stanley, Jennifer Lynch, and spy blimp technology: http://www.wash-ingtonpost.com/business/technology/blimplike-surveillance-crafts-set-to-deploy-over-maryland-heighten-privacy-concerns/2014/01/22/71a48796-7ca1-11e3-95c6-0a7aa80874bc_story.html

Julian Assange on being part of the state: http://rt.com/news/assange-surveillance-nsa-total-698/

Gallagher and Greenwall: https://firstlook.org/theintercept/arti-cle/2014/03/12/nsa-plans-infect-millions-computers-malware/

Backdoor implants: https://firstlook.org/theintercept/article/2014/03/12/nsa-plans-infect-millions-computers-malware/

SIGINT is cool: http://www.infowars.com/nsa-hacker-brags-about-how-cool-and-awesome-spying-on-innocent-people-is/

NSA Terrorist efforts minimal: http://www.newamerica.net/publications/policy/do_nsas_bulk_surveillance_programs_stop_terrorists

EU Court of Justice and http://www.commondreams.org/head-line/2014/04/08-6

Angela Merkel gets no answers: http://www.theguardian.com/world/2014/apr/10/angela-merkel-denied-access-nsa-file

Gabriel Weinberg: http://www.businessinsider.com/search-engine-duckduckgo-is-taking-on-google-by-doing-the-one-thing-they-wont-do-2014-4#ixzz2yc7FhySz

Fingerprints for lunch money: http://www.cbsnews.com/news/fingerprints-pay-for-school-lunch/

China drops fingerprinting: http://www.theregister.co.uk/2006/11/09/hongkong_kiddyprinting/

Big Brother watch and thirteen-year-old: http://rt.com/news/uk-school-children-fingerprinted-461/

Three-year-olds fingerprinted: http://www.trendhunter.com/trends/3-year-old-school-children-fingerprinted

Brian W. Walsh on unreasonable federal prosecutors: http://dailysignal.com/2011/08/08/the-worst-thing-that-anybody-can-do-to-you-is-take-away-your-freedom/

Jeff Sherwood on trying to explain the law to a judge: http://www.examiner.com/article/exclusive-man-found-guilty-of-electronic-cigarette-law-that-does-not-exist

Electronic concentration camp: http://www.lewrockwell.com/2014/01/john-w-whitehead/life-in-the-electronic-concentration-camp/

Chuck Baldwin on nonpartisan election races: http://chuckbaldwinlive.com/Articles/tabid/109/ID/1216/Police-Abuse-Riots-and-AR-15-Rifles.aspx

Randy Rider on police held to higher standard: http://www.officer.com/article/10249914/civilian-review-boards

16. THE MILITARIZATION OF POLICE:

Thomas Lifson on some victims more important that others. : http://americanthinker.com/blog/2014/08/nonwhite_cop_kills_unarmed_white_youth_national_media_ag_and_potus_ignore.html#ixzz3BQRDe4hJ

Matt Apuzzo on police departments looking like military units: http://www.nytimes.com/2014/06/09/us/war-gear-flows-to-police-departments.html?_r=0

Neenah councilman William Pollnow asks why: http://www.nytimes.com/2014/06/09/us/war-gear-flows-to-police-departments.html?_r=0

Brennan Griffin on criticism of campus police militarization: http://www.rawstory.com/rs/2014/09/05/texas-school-districts-militarize-campus-cops-with-free-surplus-weapons-armored-vehicles/

Detective John Baeza on militarized police warrior: http://www.huffingtonpost.com/2014/06/09/military-equipment-used-by-police_n_5475382.html

William Norman Grigg on police calls as military engagements: http://www.lewrockwell.com/lrc-blog/use-it-or-lose-it-federally-subsidized-police-escalation/

Daniel Rivero, Jorge Rivas, and Tim Lynch on missing 1033 weapons: http://fusion.net/leadership/story/americas-police-departments-lose-loads-military-issued-weapons-984250

Jake Tapper on comparing Ferguson to Afghanistan: http://www.nationaljournal.com/domesticpolicy/cnn-anchor-on-police-in-ferguson-monday-night-this-doesn-t-make-any-sense-20140818

Frederick Reese on Elaborated Social Identity Model: http://www.mintpressnews.com/benefits-police-militarization/195690/

Jerrail Taylor on brother's shooting: http://www.inquisitr.com/1412236/dillon-taylor-police-shooting/#ZzEt0WOdWJippx0t.99

Police Chief Chris Burbank on not releasing shooting video: http://www.inquisitr.com/1421915/dillon-taylor-shooting-camera/

IACP report on percentage of police force: http://www.usatoday.com/story/news/nation/2014/08/14/police-killings-data/14060357/

Charles C. W. Cooke on police gearing up for invasion: http://www.nationalreview.com/article/381446/barney-fife-meets-delta-force-charles-c-w-cooke

Tiger Parsons and Ronald Teachman on armored vehicles: http://www.nytimes.com/2014/06/09/us/war-gear-flows-to-police-departments.html?_r=0

17. THE RISE OF SWAT TEAMS

Radley Balko on public has no right to know: http://www.washingtonpost.com/news/the-watch/wp/2014/06/24/new-aclu-report-takes-a-snapshot-of-police-militarization-in-the-united-states/

SWAT teams increase: http://cops.usdoj.gov/html/dispatch/12-2013/will_the_growing_militarization_of_our_police_doom_community_policing.asp

Kara Dansky on escalating risk of violence: http://www.huffingtonpost.com/2014/06/09/military-equipment-used-by-police_n_5475382.html

Jessie Rossman on can't have it both ways: http://www.washingtonpost.com/news/the-watch/wp/2014/06/26/massachusetts-swat-teams-claim-theyre-private-corporations-immune-from-open-records-laws/

Mother on Bounkham "Bou Bou" Phonesavanh hit by flash-bang grenade: http://www.salon.com/2014/06/24/a_swat_team_blew_a_hole_in_my_2_year_old_son/

Phonesavanh attorney's report: http://www.theorganicprepper.ca/cost-of-medical-bills-for-baby-hit-by-swat-grenade-over-800000-countys-refusal-to-pay-priceless-08192014#sthash.EsT8QBGM.dpuf

Daisy Luther on lack of accountability: http://www.theorganicprepper.ca/cost-of-medical-bills-for-baby-hit-by-swat-grenade-over-800000-countys-refusal-to-pay-priceless-08192014#sthash.EsT8QBGM.dpuf

Mark Witaschek on risking job loss: http://www.policestateusa.com/2013/mark-witaschek-ammunition-charge/

Mike Adams on something terribly wrong in America: http://www.naturalnews.com/045141_USDA_submachine_guns_paramilitary_agencies.html

Police Chief Timothy Fitch trained in Israel: https://twitter.com/tparsi/status/499917611715813376

Max Blumenthal on "Israelification" of U.S. police forces: http://exiledonline.com/max-blumenthal-how-israeli-occupation-forces-bahraini-monarchy-guards-trained-u-s-police-for-coordinated-crackdown-on-occupy-protests/

Police exchange program GILEE: http://sfbayview.com/2011/05/police-training-exchange-compounds-us-israeli-racism/

Eran Efrati on Israeli training of U.S. police: http://countercurrentnews.com/2014/08/former-israeli-soldiers-message-to-protesters-in-the-us-youre-next/

Police lieutenant Davidson knocks over wheelchair: http://fox59.com/2014/07/02/video-lafayette-officer-keeps-job-after-pushing-over-man-in-wheelchair/#axzz36qNNLQhC

John Whitehead on private police: http://www.washingtonsblog.com/2015/03/private-police-mercenaries-american-police-state.html

18. DEADLY FORCE

D. Brian Burghart on intentional suppression of police killings: http://gawker.com/what-ive-learned-from-two-years-collecting-data-on-poli-1625472836

Kevin Johnson, Meghan Hoyer, and Brad Heath on flawed database of police shootings: http://www.usatoday.com/story/news/nation/2014/08/14/police-killings-data/14060357/

Geoff Alpert on no national database: http://www.usatoday.com/story/news/nation/2014/08/14/police-killings-data/14060357/

Douglas Zerby's mother on police shooting: http://thefreethoughtproject.com/police-shoot-kill-man-watering-lawn-family-awarded-6-5-million/

Christian Alberto Sierra shooting: http://www.washingtonpost.com/local/crime/teen-shot-by-purcellville-officer-was-depressed-friend-had-called-for-help/2014/05/27/c5d350bc-e5e3-11e3-8f90-73e071f3d637_story.html

Miriam Carey shooting: http://mobile.wnd.com/2014/04/new-revelations-about-mom-killed-by-capitol-cops/

The shooting of Daniel Saenz: http://www.elpasotimes.com/latestnews/ci_25970120/video-el-paso-police-fatal-shooting-daniel-saenz

Natasha Lennard on no collective rage: https://news.vice.com/article/el-paso-releases-video-of-cop-executing-handcuffed-man-wheres-the-anger

William Norman Grigg on state-licensed aggression: http://www.lewrockwell.com/lrc-blog/in-florida-non-submission-to-a-police-beating-is-attempted-murder/

Paul Craig Roberts on gratuitous police violence: http://www.paulcraigroberts.org/2014/08/18/militarization-police-produced-murder-machine-paul-craig-roberts/

Ken Ellis and DOJ report: http://www.theguardian.com/world/2014/may/10/albuquerque-police-shootings-violence-breaking-bad

Chief Gorden Eden's text message: http://www.koat.com/news/albuquerque-police-chief-forbids-more-doj-meetings/26417712#ixzz34H73n6MO

James Boyd shooting in Albuquerque: http://www.huffingtonpost.com/

2014/06/29/james-boyd-shooting-lawsuit-albuquerque-police_n_5541105. html

Rob Kall on plausible deniability: http://www.opednews.com/articles/ What-Comes-After-It-s-Oka-by-Rob-Kall-Extra-judicial-Killings_Ndaa-National-Defense-Authorization-Act_Targeted-Killing-140608-219.html

Captain Nicolas Aquino on officer wanting to Taze him: http://reason. com/reasontv/2014/05/22/nicolas-aquino-story

Cop shoots at teenagers: http://www.rawstory.com/rs/2014/05/30/campus-cop-fires-gun-at-teens-he-caught-making-out-in-car-parked-at-ok-school/

19. WRONGFUL ARRESTS

Radley Balko on more humane policy: http://www.washingtonpost.com/ news/the-watch/wp/2014/03/17/county-settles-with-mother-who-lost-newborn-after-erroneous-drug-test/

Peggy Duran on guilty until proven innocent: http://www.newschannel5. com/story/24811452/clarksville-student-suspended-for-knife-in-fathers-car

Jordan Wiser on lapse of judgment: http://www.huffingtonpost. com/2014/03/10/jordan-wiser_n_4921266.html?utm_hp_ref=crime

Justice Stephen Gerald Breyer on difficulty of knowing laws: http://thefree-thoughtproject.com/you-break-the-law-every-day-without-even-knowing-it/#4GIfOQ0ypuP7S9j0.99

Ginseng seized in West Virginia: http://www.wvva.com/ story/26563681/2014/09/18/police-seize-190-pounds-of-illegally-harvested-ginseng

Adopting the tactics of criminals: http://www.policestateusa.com/2014/ sneak-and-peek-warrants/

Jonathan Turley on chilling effect on citizens: http://jonathanturley. org/2014/07/14/texas-jury-refuses-to-indict-man-who-shot-and-killed-officer-during-no-knock-raid/

Chuck Baldwin on nonpartisan election races: http://chuckbaldwinlive.com/ Articles/tabid/109/ID/1216/Police-Abuse-Riots-and-AR-15-Rifles.aspx

20. FINANCE CAPITALISM

American workers receiving SSDI checks: http://news.investors. com/061714-704981-11-million-on-disability-why-are-numbers-climbing-so-fast.htm#ixzz35xeVvCUp

Richard Burkhauser on encouraging long-term unemployment: http:// www.realclearmarkets.com/articles/2013/05/15/americas_growing_ social_security_disability_problem_100322.html

Carroll Quigley on powers of financial capitalism: http://quotes.liberty-tree.ca/quotes_by/carroll+quigley

Exchange Stabilization Fund description: http://www.newyorkfed.org/ aboutthefed/fedpoint/fed14.html

Dr. Ron Paul on ESF slush fund: http://www.forbes.com/2010/01/09/ federal-reserve-audit-intelligent-investing-ron-paul.html

Treasury secretary need not comply with U.S. laws: C. Randall Henning, *The Exchange Stabilization Fund: Slush Money or War Chest?* (Washington, D.C.: Institute for International Economics, 1999), p. 12.

Lawrence Houston on un-vouchered funds: http://www.marketskeptics.
com/2010/09/exchange-stabilization-fund-role-in.html

ESF in various scandals: http://www.marketskeptics.com/2011/06/the-esf-
and-its-history.html

Marilyn Barnewall on central banker control: http://www.newswithviews.
com/Barnewall/marilyn103.htm

Mayor John Hylan on banker takeover: http://www.washingtonsblog.
com/2012/03/nyc-mayor-john-hylan-house-banking-committee-chairs-
on-monetary-reform.html

Montagu Norman on successfully accomplished plan: https://forumnews.
wordpress.com/2011/04/04/the-ultimate-bankster-quote-from-a-former-
governor-of-the-bank-of-england-1920-1944/

Larry Flynt on government by corporations: http://www.huffingtonpost.
com/larry-flynt/common-sense-2009_b_264706.html

George Carlin on what they want: http://www.rense.com/general82/car-
rlin.htm

J. R. Dunn on true reason is power: http://www.americanthinker.
com/2014/01/the_progressive_agenda_crashes_into_complexity.html#ix-
zz2pednQYPE

John DiNardo on veiled world of horrors: http://thelightofdayradioshow.
com/archives/Dinardo/Elites_Nazi

Alex Jones on "scientific dictatorship": https://www.youtube.com/
watch?v=vxqPDoYUDGk

George Sugihara and the 2011 Swiss study of bank control: http://www.
newscientist.com/article/mg21228354.500-revealed--the-capitalist-
network-that-runs-the-world.html#.VLrVMskXFmo

21. DEATH OF THE SPECIES

Dr. Shanna Swan: http://www.thedailybeast.com/articles/2011/01/04/
low-sperm-count-why-male-fertility-is-falling.html

Richard Sharpe on the reality of falling sperm counts: http://www.tele-
graph.co.uk/health/healthnews/9722963/Male-fertility-under-threat-as-
average-sperm-counts-drop.html

Higher-education   costs: http://www.washingtonpost.com/blogs/
post-partisan/wp/2014/04/03/the-high-price-of-middle-class-
membership/?wpisrc=nl_popns

Work   for   schooling: http://www.theatlantic.com/education/ar-
chive/2014/04/the-myth-of-working-your-way-through-college/359735/

Jon Rappoport on linking individual and power: https://jonrappoport.
wordpress.com/2015/01/14/thought-controlled-classroom-orgy-of-the-
group/

A National Call: Save Civilian Public Education: https://sites.google.com/
site/acalltoconfrontmilitarization/home/

Tom Allon on a paradigm shift in education: http://www.huffingtonpost.
com/tom-allon/education-reform_b_1428500.html

Dennis Van Roekel on education makes America strong: http://www.nea.
org/grants/52412.htm

Shayna A. Pitre on common aspects of foreign teachers: http://www.huffington-
post.com/shayna-a-pitre/eduction-reformwhat-we-ca_b_5374817.html

22. DEATH OF THE MASS MEDIA

Michael Crichton on junk media: http://www.slate.com/articles/news_and_politics/press_box/2008/05/michael_crichton_vindicated.html

Timothy McGaw on print models uprooted: http://www.crainscleveland.com/staff/11/timothy-magaw

Crichton on media zealotry: http://www.slate.com/articles/news_and_politics/press_box/2008/05/michael_crichton_vindicated.2.html

Richard Greenfield on no rapid change in TV viewing: http://www.nytimes.com/2013/01/07/business/media/for-legacy-media-companies-a-lucrative-year.html?src=dayp&_r=0

Jack Shafer on Crichton's observations: http://www.slate.com/articles/news_and_politics/press_box/2008/05/michael_crichton_vindicated.html

Hadas Gold on older TV viewers: http://www.politico.com/blogs/media/2014/05/may-cable-news-ratings-spare-no-one-189393.html

Newspaper revenues and circulation declining: http://stateofthemedia.org/2013/newspapers-stabilizing-but-still-threatened/12-newsroom-workforce-still-dropping/

2013 Gallup poll on TV trust: http://www.gallup.com/poll/163052/americans-confidence-congress-falls-lowest-record.aspx

Joshua Krause on ratings based on truth: http://www.thedailysheeple.com/death-throes-the-mainstream-media-is-hanging-by-a-thread_062014#sthash.O7egB6DA.dpufDeath

Alternative media on levels of authenticity: http://www.infowars.com/alternative-media-upstages-lamestream-media-in-world-class-coverage-of-historic-bundy-ranch-showdown/

Anthony Gucciardi on kicking out mainstream media: http://www.storyleak.com/gallup-poll-virtually-no-one-trusts-the-mainstream-media/#ixzz33tdTgl7p

David Brock and Paul Joseph Watson on MMFA as attack dog for Obama administration: http://www.infowars.com/media-matters-boss-admit-soros-funded-group-works-to-destroy-alternative-media/

Michael Rivero on propaganda gives excuse: http://quotes.libertytree.ca/quotes_by/michael+rivero

Alain de Botton on ideal news organization of the future: http://theweek.com/article/index/256737/the-future-of-news

Clay Shirky on only partisans left: http://www.politico.com/magazine/story/2014/02/tv-is-dead-now-what-103670.html#ixzz3PI7QvYWQ

2014 Gallup poll on media trust: http://www.gallup.com/poll/176042/trust-mass-media-returns-time-low.aspx

Elliot D. Cohen on not allowing ourselves to be suckered: http://www.projectcensored.org/digging-deeper-politico-corporate-media-manipulation-critical-thinking-democracy/

Michael Parenti on the world as a scatter of events: http://www.michaelparenti.org/MonopolyMedia.html

23. COMING COLLAPSE?

Financial experts forecast hard times and collapse: http://theeconomiccollapseblog.com/archives/dent-faber-celente-maloney-rogers-what-do-they-say-is-coming-in-2014

Mike Adams on inevitability of collapse: http://www.naturalnews.com/046075_Health_Ranger_poverty_financial_survival.html#ixzz382gqWYws

Jack Curtis on America tending to its own business: http://www.americanthinker.com/2013/03/bread_and_circuses_the_last_days_of_the_american_empire.html

Michael T. Snyder on Chinese acquisition of U.S. businesses: http://theeconomiccollapseblog.com/archives/meet-your-new-boss-buying-large-employers-will-enable-china-to-dominate-1000s-of-u-s-communities

Davvy Kidd on social breakdown: http://www.newswithviews.com/Devvy/kidd644.htm

Brandon Smith on final swindle of American wealth: http://www.alt-market.com/articles/1977-the-final-swindle-of-private-american-wealth-has-begun

NASA-grant study on fall of civilizations: http://www.alternet.org/economy/nasa-funded-study-industrial-civilization-headed-irreversible-collapse-due-inequality

Nafeez Ahmed and "HANDY" study: http://www.theguardian.com/environment/earth-insight/2014/mar/14/nasa-civilisation-irreversible-collapse-study-scientists

Brian Brown on mounting water crisis: http://www.nbcnews.com/news/us-news/last-drop-americas-breadbasket-faces-dire-water-crisis-n146836

Mayors Conference survey: http://www.governing.com/topics/health-human-services/gov-cities-report-homeless-hunger-increase-fewer-resources.html

Brandon Smith on second American Revolution: http://www.alt-market.com/articles/2180-a-second-american-revolution-is-now-inevitable

Henry Giroux and John W. Whitehead on trusting government: http://www.lewrockwell.com/2014/06/john-w-whitehead/you-trust-the-government/

The Minerva Initiative projects: http://minerva.dtic.mil/funded.html

David Price and American Anthropological Association concerns: http://www.rawstory.com/rs/2014/06/12/defense-dept-studying-protesters-to-prepare-for-mass-civil-breakdown/

John W. Whitehead on arms and ammunition for federal agencies: http://www.lewrockwell.com/2014/06/john-w-whitehead/you-trust-the-government/

Witnesses to looting of wrecked car: http://www.chron.com/news/houston-texas/houston/article/Onlookers-allegedly-steal-groceries-from-woman-5531076.php

Mike Adams on normalcy bias: http://www.naturalnews.com/045465_stealing_groceries_fatal_crash_food_riots.html#ixzz33zdB5oug

Alain de Botton on knowing what to do with facts: http://theweek.com/article/index/256737/the-future-of-news

Richard Scheck on how to live with the Leviathan: http://jimmarrs.com/news_events/news/living-with-the-leviathan/

Allen West and the six conundrums: http://allenbwest.com/2014/05/9-conundrums-left-answer/

Jules Dervaes and his microfarm: http://www.naturalcuresnotmedicine.com/2013/07/1-man-produced-6000-pounds-of-food-on.html

FEMA on preparedness basic kit: http://www.ready.gov/build-a-kit

One hundred websites with e-books: http://www.naturalcuresnotmedicine.com/2014/09/free-books100-legal-sites-download-literature.html

Robert David Steele on Open Source Agency and bottom-up action: http://www.theguardian.com/environment/earth-insight/2014/jun/19/opensource-revolution-conquer-one-percent-cia-spy

Paul Rosenberg on dropping political ideologies: http://ncrenegade.com/editorial/political-ideologies-politics-makes-life-ugly/

Joseph Stiglitz on Iceland did the right thing: http://www.bloomberg.com/news/2011-02-01/iceland-proves-ireland-did-wrong-things-saving-banksinstead-of-taxpayer.html

Paul Krugman on Iceland's heterodoxy: http://krugman.blogs.nytimes.com/2010/11/24/lands-of-ice-and-ire/?_php=true&_type=blogs&_r=0

CNN *iReport* on why no media reports on Iceland: http://ireport.cnn.com/docs/DOC-787377

More American receive benefits than paid in taxes: http://taxfoundation.org/article/accounting-what-families-pay-taxes-and-what-they-receivegovernment-spending-0

Americans on food stamps and unemployment: http://www.statisticbrain.com/welfare-statistics/

Alfred W. Evans on "put me in charge": http://www.wacotrib.com/opinion/letters_to_editor/letters-fixing-social-security-put-me-in-charge-ofwelfare/article_7fc4324b-9fd3-52b3-8c5c-2ecd2c1c862a.html

Robert Hase with Elias Alias on highest law enforcement authority: http://oathkeepers.org/oath/2014/09/08/sheriff-nace-stands-tall-for-gun-rightsin-pennsylvania/

Andrew Martin on creating realities of delusion: http://www.collectiveevolution.com/2014/08/02/this-is-what-the-world-needs-now-in-order-tochange/

Grazyna Fosar and Franz Bludorf on changing DNA through thought and language: http://wakeup-world.com/2011/07/12/scientist-prove-dna-canbe-reprogrammed-by-words-frequencies/

Dr. Andrew Newberg and Robert Hummer on religion and spirituality enmeshed: http://content.time.com/time/magazine/article/0,9171,1879179,00.html

INDEX

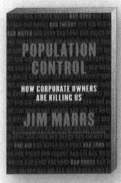
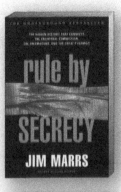